WHERE ARE THEY?

Steven Lazaroff & Mark Rodger

RodgerLaz Publishing S.E.N.C.
2475 Palmipedes
St Laurent, QC, Canada, H4R 0J2
www.Rodgerlaz.com

Ordering Information:
Quantity sales. Special discounts are available on quantity purchases by corporations, associations, and others. For details, contact the publisher at the address above.
Orders by U.S. trade bookstores and wholesalers. Please contact the publisher at the address above or by email: slazaroff@rodgerlaz.com

Printed in Canada and the United States

Publisher's Cataloging-in-Publication data
Steven Lazaroff and Mark Rodger.
Where are they? / Steven Lazaroff and Mark Rodger
p. cm.
ISBN 978-1-7752921-4-2
1. The main category of the book —Cosmology, Astrophysics & Space Science, Astronomy,

First Edition

14 13 12 11 10 / 10 9 8 7 6 5 4 3 2

CONTENTS

ACKNOWLEDGMENTS

We would like to acknowledge several people.

First and foremost, to our significant others, who put up with this project. The amount of time spent on this project took away from the time we should have been paying attention to you both.

We want to thank out beta readers, bloggers and friends that have taken the time to review this work and help us with their thoughts and constructive criticism. The list of names is too long to cite here.

To John, thanks for all your help with this book. You have our eternal gratitude. To Marco Lambert and Daniel Lefaivre for the editing on the French version, merci beaucoup!

Finally, to friends and family who supported us and promoted us with our first book, we hope to count on them for this work and start the process over once again.

Introduction

Three statements:

- 100 million planets scattered in galaxies throughout the universe are home to life, have been home to life in the past, or will be home to life at some point in the future.

- There is no life on any planet other than Earth; there never has been and there never will be. We spin through the universe alone. When Earth reaches its unavoidable end, life will have been extinguished, everywhere and for ever.

- There is life on one other planet. It's so far away from us that we will never be able to make contact, never mind visit. What form does this life take? Is it intelligent? If so, does it look like us or is it so utterly different that we would be appalled merely to look at it? We don't know, and we never will.

Three possibilities. And we don't have a clue which (if any) is correct. No wonder the search for life on other planets and speculation about what it would be like has so many in its grip! We've been looking, and looking, and looking – and we haven't found a trace.

Does that mean there's nothing there? Or does it mean that we're not looking in the right way, or the right place, or we don't yet have the technology for effective search? If life – intelligent life – exists elsewhere, where is it? Why have we not found it, and why has it not found us?

(And that's a big question. How do we know it hasn't found us? A lot of people believe it has and that we simply didn't notice, either because it came in disguise or because we didn't know what to look for).

Two more statements:

- There is life elsewhere in the universe, but it has not yet reached our stage of development. Only on Earth has civilisation developed.

- Our civilisation is a poor thing when compared with some that exist. True, there are planets in which intelligent life still lives in caves, lighting fires to keep marauding animals away, but there are others that are more advanced than ours by half a million years or more. They have technology, energy sources, materials, communication channels and so many other things that we don't yet even dream of.

The first of those two statements – would anyone really want that to be true? Go to a hardware store this Saturday morning and watch the DIYers emerging with tools and materials for the weekend's tasks. Would it be inspiring to think that this was the highest expression of civilisation anywhere in the universe? Or might it prove depressing?

But now turn to the second statement. Those lifeforms that are 500,000 years ahead of Earth – if they come here, what might they be capable of? Could they – would they want to – enslave humans? Kill them for sport? Keep people as pets to make interesting dinner party conversation subjects?

The possibility of extra-terrestrial life has gripped humans for as long as anyone knows. *The War of the Worlds*, encouraged by the idea (now known to be false) that there were artificial canals on Mars, placed life as close as that. And now we know that Mars has permafrost. Permafrost means water. Water means the possibility of life. There's no expectation that something resembling a person will be swimming in it. What's more likely – or, at least, possible – is some forms of bacteria.

But a bacterium is life. Its existence would mean that the conditions for life to develop had existed elsewhere than on Earth. And if bacteria can develop, then who's to say that there isn't, somewhere, a soccer team competing for another planet's World Cup?

In fact, bacteria on Mars wouldn't answer the big question. Rocks disturbed by huge meteor and asteroid impacts have travelled between the two planets. There's nothing to say that bacteria on Mars did not get there from Earth. Come to that, there's nothing to say that intelligent life did not develop on Earth as a result of primitive bacteria that travelled from Mars and found better conditions for development here. Because the big question is: has life developed from scratch in more than one place in the universe?

It's a question scientists are wary of. Charles Darwin, when he mapped out the way that life forms evolved, carefully avoided any mention of where they came from. He didn't talk about God, and nor did he talk about blind chance. There are scientists who will say that even the simplest lifeforms are so complex that it is almost impossible that the sequence of processes that led to formation of the first bacterium was ever reproduced elsewhere in the universe, and so we are alone. And there are others who say, 'If it's that complex, how could it have happened by accident? And if it was planned, by whom? And by what stretch of arrogance would earthlings claim that it was planned only once, and uniquely to this planet?

Fred Hoyle and N. Chandra Wickramasinghe said that it was impossible that life could have started by chance. The odds, they said, amounted to one chance in 10 to the power of 40,000. Since most ordinary humans can't visualise such a number, they helpfully added that it was "an outrageously small probability." Richard Dawkins disagrees and treats with contempt anyone who believes that the development of life could even theoretically have been anything but an accident. Stephen Hawking, in his last book before he died, said that the universe was not created by God. Instead, it was the inevitable result of the laws of nature. Yes, Stephen. Quite so. And the laws of nature – where did they come from?

The purpose of this book is not to take one side or the other in that argument. It is to explore the present state of knowledge and to say where humanity now stands on the question of whether or not we are alone in the universe. Because there isn't the slightest doubt:

that is a question that has occupied humans since they became human, and it's a question that shows no sign of going away.

And if there are intelligent beings elsewhere in the universe – where are they?

CHAPTER 1

EARTHLINGS MEET ALIENS

WHAT WOULD HAPPEN IN A PHYSICAL MEETING BETWEEN PEOPLE ON EARTH AND REPRESENTATIVES FROM A DISTANT CIVILISATION?

The technology does not as yet exist here on earth to travel to a distant planet on which a civilisation may already be established. It follows that, if there is to be a meeting – a physical meeting, rather than communication – between humans and the residents of another planet, it will be the developed planet that visits earth and not the other way round.

What would that be like? What would the impact be?

The first impact would be physical. Unless conditions on the visitors' home planet are similar to those here, it's unlikely that visiting aliens will look anything like humans. And it has to be said that human beings do not have a very good record when it comes to dealing with people who look different from them. Different skin colour, different eye shape – those two things alone have inspired sections of Earth's deeply tribal population not only to dislike others, but to wage war against them. What likelihood is there that "people" who look nothing like "people" will be received with a peaceful welcome?

Would humans even think of a visitor of widely different appearance – say, one who looked like an insect – as a "person" at all? Even if the insect arrived in a spacecraft made from a material no-one on earth has ever come across and powered by an energy source that humans can scarcely imagine, is it not likely that its appearance would lead humans to disdain the visitors? Perhaps even attempt to eradicate them as an infestation?

And yet, there are good reasons for imagining that a visitor from elsewhere in space might take the form of an insect. A cockroach, perhaps, or a cricket. (What good reasons? Well, here on earth, there are more insects than the total population of all other animal life forms put together. The insect form does appear to be capable of almost infinite variation and development, and it is very difficult to wipe out).

Since humans unnerved by the appearance of their visitors might attempt to destroy them and be destroyed in their turn (this is, after all, an advanced civilisation that's calling), it's good to realise that the actual visitors may not be there in person. Scientists like John von Neumann have theorised about spacecraft that can reproduce themselves by mining the materials they need from a passing asteroid. The suggestion is that this technology will be needed in order to cover the unimaginable distances between galaxies (and even from one end of our own galaxy to the other). It is still little more than a talking point here on earth, but it may be old hat to an advanced interstellar civilisation. It has also been suggested that self-reproducing molecules could be used inside the human (or alien) body, so that an astronaut could last the desperately long periods of time that those distances would take to cover.

But how realistic is that? Yes, it's entirely possible that civilisations more advanced than ours have already developed a means to keep a body going in this fashion – but who would agree to it? Imagine that you are in the same position as one of those alien astronauts being tapped up for a journey to Earth from the galaxy MACS0647-JD. It's 13.3 billion light years away, so – if your civilisation has developed a form of transportation that will travel at the speed of light – the time spent on the journey is unimaginable. Would you want to do it? Leave the kids, your husband and your book club knowing that at the end of your journey you would encounter a civilisation a few hundred millennia less developed than yours. And that you couldn't get home for nearly 27 billion years at the earliest, by which time your planet would in all possibility have come to the end of its life? And that, when you arrived on Earth, your body would have been renewed some eighty times, so you wouldn't really still be you at all?

Really, the imposition on someone asked to travel that far is impossible. It's much more likely that some kind of hologram image will be in the spacecraft. Or a robot – although in that case the self-reproducing technology would still be required.

Of course, MACS0647-JD is, at least in the present state of knowledge, the furthest galaxy from Earth. There are galaxies less than a million light years from here, and it's not impossible that an interstellar astronaut would be prepared to make that journey.

And then there's the question of how such an advanced civilisation might have developed its bodies, with the related question: what is it that makes a person a person? If these people are anything like humans, their lifespans are subject to physical limitations. By the age of seventy, a human being here on Earth may have developed an enormous amount of experience and a great deal of knowledge, but that knowledge is contained in a body that is no longer what once it was. As long as the person thinks of the physical shell in which the mind is enclosed as the self, there's nothing to be done about that beyond replacement of a few incidental parts like hip joints. If, though, it is the mind itself that the person sees as "her" or "him," then there is no barrier except technology to stand in the way of replacing body parts wholesale – for example, with components made of long-lasting metals.

That technology is only beginning to exist on Earth but may be very advanced elsewhere. Couple that with developments on Earth in AI (Artificial Intelligence), picture how far ahead more advanced civilisations may be in the AI field, and it becomes possible to say that "I" is now "It." Not just the ability to reason, but knowledge and intelligence itself has been transferred into something that, at first sight, could be regarded by someone on Earth as some kind of machine but that regards itself as a person.

IF THE FIRST CONTACT IS NOT PHYSICAL BUT BY MESSAGE

It's probably much more likely that the first "meeting" between extra-terrestrial beings and people on Earth will be the receipt of a transmission from a faraway planet and not a physical meeting. But how will that message be received? Here are two scenarios:

- Watchers, whether at SETI or elsewhere, pick up signals in a form that gives a high degree of certainty that they were transmitted with purpose by intelligent beings in the hope that they would be picked up by some other intelligent beings somewhere, sometime. The signals cannot be translated as having any particular meaning; they simply amount to an indication of intelligent origin.

- A message is received in Chinese, Spanish, English, Hindi, and Arabic. The message begins: "Hello, Earth."

Clearly, the impact of these two contacts will be very different. The first says that humans are not alone in the universe. There is other life out there, and it's intelligent enough to be able to transmit signals deep into space. Whoever sent those signals sent them at random in the hope that, somewhere, there would be other intelligent beings capable of receiving them and understanding that they had been sent deliberately. They amount, in effect, to, "Are we alone? Can anyone hear us?" But the second – ah, the second! That message says, "We know who – and where – you are." The choice of languages indicates that the sender has studied Earth at least well enough to know the five most common native languages. However far away they may be, they have the technology to be able to say, "We're watching you."

How would people on earth respond to either of these contacts? It depends to a large extent on who receives them. Would it be a government? A member of the public? Or both?

First, the undirected random signal. If this is picked up by a government, it would be good to think that knowledge of its arrival would be shared with other governments and with the public. Good – but perhaps naïve. Governments tend to tell their voters as little as possible. In this case, they might say they were keeping the information secret because they did not wish to see people unnecessarily frightened. It's a good story; probably, though, it would be more true to say that they saw a benefit in having information that other people did not have. It's also likely that any government would see profit in becoming the conduit through which the extra-terrestrial civilisation deals with Earth. That would be a lot easier if, at least at the start, only they knew about it.

If the signal was first received by a member of the public – and that is by no means out of the question, because a great many recent astronomical discoveries have been made by amateurs – then a variety of possibilities emerges. The individual might decide to keep the discovery secret but to offer it to a government (or some other agency) for money. That's unlikely, though, given the kind of people who carry out this sort of observation. They are much more likely to want to tell the world about it. The news would be published in specialist magazines and rapidly picked up by the wider press. Governments then would have two options: to endorse the news already in people's possession and try to influence the way it is handled; or to ridicule the member of the public and say that the information being spread around was untrue. In some jurisdictions, the observer would probably be dispatched to an asylum for the differently sane.

What is most likely, of course, is that the signal would be received almost simultaneously by more than one government (they're all looking, whether they publicise the fact or not) and by more than one member of the public in more than one country. In that case, controlling the information would become impossible and the impact would be influenced by how the public reacted. That reaction would also influence Earth's response – would there be one? If so, what form would it take? A lot of thought has already gone into establishing how to communicate information to an alien civilisation, in much the same way as the Rosetta Stone made it possible for the modern world to interpret the hieroglyphics used in ancient Egypt. (And that is a good example to use, because the Egyptian civilisation that produced the Rosetta Stone was dead long before it was translated – and the extra-terrestrial civilisation that sent the signal may also have expired before it is received on Earth).

And now the signal that began, "Hello, Earth." The same questions have to be asked concerning who receives the message and what they do with it, but the impact for human civilisation is very different. How does the message continue? Perhaps it says, "This message informs you that you will soon receive a visit from a spacecraft containing five of our citizens." Earth has been put on notice that ET is about to become a reality – Earth is to receive visitors from outer space. If the transmitting civilisation has any sense (and would it, otherwise, have sent this message?), it will also have sent pictures of the aliens so that they would be recognised on arrival, given the appropriate respect, and not (if they do happen to look like insects) sprayed with bug-killer.

There are many possible continuations to the message.

Here are two:

- "Our intention is to colonise the Earth. No one will be hurt, as long as you accept our rule. Do not attempt to resist. We have weapons that you cannot begin to imagine; we would prefer not to have to use them, but we will do so if you make it necessary."

- "We come in peace. We offer friendship. We believe we have technology to share that you will wish to benefit from."

Those two options will bear closer inspection.

COLONISATION

"We have weapons that you cannot begin to imagine." They did not exaggerate. They had the technology to build a spacecraft capable of withstanding the stresses and pressures of getting from there to here at a speed approaching that of light. When they landed, they did so in public – not in some remote desert, as the movies would suggest, but in a park in the centre of a large and crowded capital city. There was no chance that their arrival would not be noticed, and none that it would be kept secret. The country in which they had landed did not dare to attack the spacecraft with nuclear weapons, for fear of what that would do to its own citizens and seat of government. Other countries saw no such reason for restraint and attempted to unleash a nuclear attack.

The result came as a shock. The invader had the ability to suppress all earth-bound communications. Rockets could not be launched, troops could not be mobilised, and nuclear devices could not be exploded. Soldiers who attempted to resist invasion received orders (that appeared to come from their own top brass) to turn their guns on each other. Thought-management overcame any resistance they might offer to those orders.

An army of fighting robots was assembled from nanotechnology or some similar technique. Humans had no way of stopping or even slowing down those robots – but the robots could kill humans in the blink of an eye. In short, a civilisation that was capable of reaching Earth and came with malign intent was not going to be defeated by any force that Earth could muster. It would have been like stone age warriors fighting against a modern army, air force and navy by throwing rocks.

So surrender was the only option? Well, surrender is never the only option, but that is only true for people who are prepared to die for a lost cause. History is full of people like that. They are remembered with the reverence due to heroes. Unlike the people who stayed behind, they did not become slaves. But those heroes are always in the minority, because most people value staying alive above all other possibilities. And so, yes, surrender was – for almost everyone – the only option.

But that was a brute force invasion. Not everyone may know they have surrendered. A great deal depends on the form of the colonisation. It's possible to describe colonisations where almost no one need know that an invasion has taken place at all. A civilisation of the kind imagined here may be capable of replacing Earth's existing leaders with people who look exactly like them but who are, in fact, aliens who have assumed exactly the same body shape, appearance and speech patterns. Alternatively, they can exercise total mind control over the leadership. And, if they can do that, they can also control the minds of the people and achieve what earthly politicians have never managed: total public agreement with their policies.

A great deal will depend on the purpose of the colonisation. Is it for trade? That may seem unlikely at first glance, but the relatively developed European civilisations of the 16th, 17th and 18th centuries found trade a very profitable exercise with the backward civilisations of Africa and the South Pacific, as well as rather more advanced lands in Asia. It's possible to argue that, if the aliens want to trade with Earth, they would need to treat people here well; unfortunately, the experience of those same people colonised by Europeans suggests that that is not necessarily so.

And those trade arrangements of the past were very one-sided. If our experience with extra-terrestrials mirrors that of peoples colonised by Europeans, "trade" will not be something conducted by equals – Earth will have resources that the aliens need and what they will offer in return will amount to the equivalent of glass beads. But then, it's as well to remember that the kind of civilisation imagined here is likely to regard an iPhone X as the equivalent of glass beads.

A close look at European colonial history suggests other possibilities, too. Earth may be invaded to spread religion, as many countries on this planet were invaded to force people to adopt Christianity. Or Islam. Or Communism. And that brings us back to the question of how people will respond. This will be dealt with in more detail later in this book, but it is often suggested that one of the big losers on Earth in the event of discovery of an extra-terrestrial civilisation would be religion. And that is not, in fact, necessarily so. A cartoon doing the Facebook rounds recently showed some aliens on another planet talking to visitors from Earth. 'Jesus?' says one of the aliens. 'We know him well. He drops in about once a month. We give him coffee and cakes. How did you treat him?' Suppose the invaders brought with them a holy book that told roughly the same story as the Bible. Or the Quran. Or some other religious text already known here. Is it possible to imagine anything that could better reinforce an existing religious message?

Then again, there's the possibility that the extra-terrestrials' values are so far from those of humans that it is impossible for humans to understand them. Earth provides a fairly benign environment for the human race to grow up in. When humans have been able to desist from killing each other, in the cause of religion or trade or simply because humans are tribal and don't really like people from other tribes, the world has mostly been kind to them. Imagine a different planet – the one from which the inter-stellar visitors are coming – in which every day is a struggle for survival. These aliens did not emerge unscathed from their own planet's evolution. They survived because they became better at killing other species than other species became at killing them. What mindset – what value system – are such people likely to have developed?

It's extremely likely to be the idea that the universe is predicated on the survival of the fittest. They believe that, if humans are unable to resist them effectively, humans are not fit for survival. They would be justified in killing us, just because they could. Justified not, perhaps, by our standards and values, but certainly by theirs – and it would be their standards and values that counted, because "might is right."

THE ALIENS COME IN PEACE

It's possible to imagine contact with extract-terrestrials, either by visit or through communication, in which the aliens' intention is altruistic. They know we're here, they see us struggle, they are hundreds of thousands of years ahead of us in their civilisation, and they want to help. That must, surely, be a good thing?

Well, perhaps it is. And then again, perhaps it isn't.

The visitors look at us. What we do. How we live. They see the things we could do better. We are still using destructive and resource-depleting building materials and methods. We have not developed to anything like their level in the power to communicate with each other. Our wasteful use of energy makes them despair. Don't we understand that energy is everywhere, that it's freely accessible, and that it can be available to everyone? When it comes to healthcare, despair does not begin to describe their feelings. How can we possibly allow illness, incapacity and disease to be rife like this? Don't we understand how much better everyone's life could be? That it isn't necessary to face old age in a state of decrepitude? That disability, whether physical or mental, is easily repaired? Well, they reason, we can't understand it, can we? Because, if we did understand it, we'd change.

But is that really true?

We know we shouldn't smoke – but huge numbers of us still do. We know that building more roads and more cars is not really the most sensible way of getting around – but it's still the one that almost everyone prefers. The means exist today for every human being on this planet to enjoy a living diet and have access to clean water. We can do it. But we don't. Why not? It has to be because we choose not to. And, if we can't make the most effective use of the technology we already have, what possible reason is there to imagine that we would be better with a technology we don't yet possess?

But now we are in contact with an altruistic alien civilisation far advanced from ours. We'll learn from them. Right?

Well... It might be instructive to take a close look at what happens to aid sent by well off, developed countries to poorer, less developed countries. This is not the place to launch an attack on the elites running a number of Third World countries, but there is a great deal of evidence to say that more than half of the money sent to relieve the poorest people in those countries never reaches them. Will the fact that the aid is now coming from somewhere in outer space change things? Or is it not at least as likely that now we will see the elites of the richest countries also benefiting while those at the bottom of the pile see very little change in their lives?

The fact is that tribalism is not the only drawback affecting the human race. Or not, at least, if a tribe is a group of people from elsewhere. Tribes exist within countries. People divide themselves from others on the basis of religion, colour, accent, education, class (which may or may not embrace any or all of those factors) and other things. That seems to be the human way of doing things: to form small units isolated from others. It may be about fear. And, if it is, things are unlikely to be improved by the fact that a more advanced civilisation is offering more advanced technology, because some of the worst wars in history have been caused by feelings of inferiority. Humans do not like to feel inferior. It makes them lash out. And inferiority will be inescapable in the context of an extra-terrestrial technology transfer.

There is no reason to believe that all that will change for the better the moment someone lands from another planet and says, "Let me show you how to get unlimited energy at no cost." It would be nice to think that everyone in the whole world would immediately benefit from that introduction, but it's rather more likely that some factions would seek to take the benefit for themselves and sell to other people at a price the energy for which they paid nothing.

And what happens if the visitors bypass all the predators waiting to seize on this new technology and keep the benefits for themselves? If – as they would probably be capable of doing – they make it available to everyone. What happens? Well, clearly, the answer could be: some very good things indeed. In places where the water was salt, or polluted, it is now clean, fresh, and pure. The housing now being built is cheap, long lasting, but functional and beautiful. Food is now being produced in places that not long ago were desert.

Freedom! Liberation! At long last, the human race has the freedom it has longed for as long as stories have been told and written down. Does this mean the spread of universal human happiness?

It would be nice to think so. But we have overlooked government, and the reasons people go into it. We have taken away the ability of one set of people to exercise dominion over other sets of people. They may be reluctant to stand for it. This may usher in a time of conflict greater than anything Planet Earth has ever seen.

The title of this book is *WHERE ARE THEY?* What the book seeks to establish is: if civilisations exist on other planets, why have we as yet seen no contact from them? And perhaps the reason is right here. They are watching us. They know the level of development we have reached. They are capable of judging when we are ready to share in the cornucopia they already enjoy.

And they know we are not there yet.

CHAPTER 2

THE SETI PROJECT

Officially, the SETI Project is the Search for Extra-terrestrial
Intelligence, but it's actually a little more tightly defined than that.

QUEST FOR FIRE (originally, in French, *LA GUÈRE DU FEU*),
was a French movie made in Scotland, in Kenya, and on Vancouver
Island and set in the Europe of 80,000 years ago. It was about control
of fire – who had it, who wanted it, and how they could get it – but it
ends in a way relevant to this chapter. The love interest is provided
by Ika, a young woman played by Rae Dawn Chong, and Naoh (not
Noah), a young man played by Everett McGill. At the very end of the
movie, the two expectant lovers (Ika is pregnant) sit on the ground in
an embrace and gaze at the moon. They speak to each other in a
language supposedly developed just for this film by British author
Anthony Burgess, though in fact he lifted it wholesale from northern
Canada's Cree/Inuit people. (The Inuit got a special kick out of the
movie because the words being spoken had nothing to do with the
on-screen action – but most viewers would not know that). Their

gestures and facial expressions make it clear that they are talking about the moon. What is it? *Where* is it – how do you get there? Are there animals that could be killed for food? Are there animals that might kill *them* for food? And are there people there? People like Ika and Naoh?

This is the conversation of intelligent humans. They are able to build on their experience, discuss it in terms that others can understand, pose questions and suggest possible answers. Intelligence is at work. If someone on another planet had a conversation like this, that would be regarded as extra-terrestrial intelligence. Anyone looking at Ika and Naoh from another galaxy would draw the same conclusion: this is extra-terrestrial intelligence.

But it is not the sort of extra-terrestrial intelligence that the SETI Project is ever going to find, because Ika and Naoh are not capable of sending signals into space. However intelligent they may be, they lack the technology to do that.

So the project's name is not entirely accurate. Really, it is the search for extra-terrestrial intelligence that has reached a stage of development at which it is capable of transmitting signals into space.

And that is inevitable, because a civilisation on another planet, perhaps in another galaxy, is most likely to come to the attention of watchers on earth if it can transmit signals that can be recognised here for what they are.

The search exists in its own right, and has done for as long as humans have been looking at the sky through telescopes. In the last 30 years or so, the SETI Institute has become one of the key searchers. The common view is that the SETI Institute is a bunch of scientists examining radio waves looking for characteristics that would say, "This is not noise. This did not happen by chance. This is a signal initiated and transmitted by an intelligent being." The reality is more complex and more diverse; the Institute has three centres, a variety of purposes and a number of routes to achieve them.

The **Carl Sagan Centre**, named after astronomer and author, Carl Sagan, has more than eighty scientists pursuing six main areas of research:

- Astronomy and Astrophysics
- Exoplanets
- Planetary exploration
- Climate and geoscience
- Astrobiology
- SETI

As that list shows, the search for extra-terrestrial intelligence is only one of the Carl Sagan Centre's research interests, and scientists at the Centre have been known to become irritated when questions focus on that area alone.

The **Centre for Education** has only a peripheral interest in extra-terrestrial intelligence; it is there to promote knowledge of astronomy and astrobiology (and, indeed, space science in general) by giving educators and students the tools, resources and information they need.

The **Centre for Public Outreach** has to be aware of SETI, because this is the part of the organisation that reaches out to the public and the public is probably more interested in SETI than in any other aspect of space research. In fact, the public outreach centre runs weekly programmes on SETI.

WHERE THEY LOOK, AND HOW THEY LOOK

The most obvious place to find intelligent extra-terrestrial life is on an exoplanet. Exoplanets are planets outside Earth's solar system that orbit a star. Scientists have been convinced for centuries that such planets must exist, but the technology to confirm that a heavenly body was indeed an exoplanet is fairly new. Even as recently as 1917, it was impossible to confirm that the body that had been sighted was an exoplanet. It wasn't till 1992 that scientists could confirm that what they thought was an exoplanet was indeed an exoplanet. Now… well, now we know that there are rather a lot of them.

Not every exoplanet is a good candidate for being a home to intelligent life. Because intelligent life cannot exist on earth without water, scientists assume that water is essential for intelligent life anywhere. They don't know that for certain – it's possible that there are intelligent, functioning organisms gathering in a bar at the very moment you read this to enjoy a glass or two of liquid hydrogen – but the search at present is for exoplanets in the Goldilocks zone, where (just like Baby Bear's porridge) it's not too hot and not too cold but just right for the formation and maintenance of liquid water. (Scientists prefer the expression "habitable zone" to Goldilocks zone; the habitable zone is an area close enough to the planet's star for water to form without being so close that it boils away or so far that it freezes).

As to search methods, these combine both optical and radio telescopes. An optical telescope shows where an exoplanet is, where it is in relation to its star, and (if the telescope is sufficiently powerful and can be brought close enough to the exoplanet) whether there is any indication that conditions suitable for life actually exist on the planet. A radio telescope does the same thing, though it requires more interpretation, but a radio telescope is also capable of picking up signals transmitted from somewhere in space, and that is the way a signal is likely to be found.

WHAT THEY LOOK FOR

Almost every kind of body in the universe transmits energy (except black holes, which absorb energy and refuse to allow it out again. But more on that later in this book). The energy is transmitted in waves and the wavelength determines how people on earth "see" the energy and what name they give it. The light waves picked up by optical telescopes and the light waves picked up by radio telescopes are on the same spectrum but at different places – it is the length of the wave that determines where the source sits on the spectrum.

A possible confusion that should have been removed by that last paragraph is between light and sound. The waves picked up by a radio telescope are not – despite the word "radio" in radio telescope – the same as sound. Radio telescopes receive a form of light wave, and not sound waves. Light can travel through a vacuum, and sound cannot (which is why, when you scream in space, no one hears). There are also differences in the shape of the wave; light waves are transverse and sound waves are longitudinal.

The fact that everything is transmitting energy, and transmitting it in every possible direction, means that the universe is a very noisy place and it is very difficult to pick out the specific kind of signal that SETI is looking for, which is signals that raise at least the possibility that they were generated by intelligence. The radio frequencies used on earth lie between 20 kHz and 300 GHz, but the whole radio spectrum is a lot bigger than that – it runs from 3 Hz at one end of the scale to 3 THz (3000 GHz) at the other. Signals being generated by heavenly bodies simply because they are heavenly bodies take up the whole of that radio spectrum. SETI's scientists think it is extremely unlikely that any intelligent being would attempt to signal over such a wide range, and what they are looking for is primarily narrow-band signals (signals covering only a small part of the radio spectrum).

They also look for brief flashes of light – and by brief they mean flashes that last nanoseconds.

Signals meeting both of these criteria – narrow-band radio spectrum signals and flashes of light lasting nanoseconds – have been discovered. What happens then is that they are considered possible candidates for intelligent signals and examined in greater depth.

Scientists are also aware that picking up a radio signal would not necessarily indicate that it had been intended for us. Some signals, if strong enough, may simply have left the planet's atmosphere and reached us although no intention existed that that should happen. The same thing may be true in reverse; somewhere out there in space may be a civilisation that is trying to extract intelligent messages from chat shows and reality TV originating on earth.

And that raises one of the problems with the SETI Project: if they succeed in identifying a signal as definitely originating from an intelligent life form, how likely is it that they will be able to decode the message? As well as looking for communications or signals beamed into space, and those seemingly intended for Earth, SETI also looks at signals passing between two worlds where the line of sight extends to this planet. Such signals may be messages between a planet and another planet or a satellite, and may continue past the intended receiver and eventually reach here. It's possible to reason that an intelligent life form capable of sending signals into space is also capable of reducing them to the barest minimum in order to make them comprehensible to a different civilisation. That won't be the case if the signal represents a message between two places sharing a common language.

WHAT STAGE HAS SETI REACHED?

Here are some interesting points from current and recent searches.

FRB121102 is, as the letters FRB in its name indicates, a fast radio burst. FRBs last for perhaps a millisecond, but the energy emitted in that millisecond can be as much as the sun has given out since the first wheat and barley were cultivated some 10,000 years ago in what was then Mesopotamia and is now Iraq. Scientists argue furiously over what causes an FRB and the fact is that no-one knows. Probably, they have a variety of causes. Possibly, one of those causes is the desire of an intelligent life form to transmit a signal.

A number of FRB's have been found; 121102 was first seen in 2012 at the Arecibo Observatory in Puerto Rico. It was seen to recur a number of times, and a team at Cornell University led by Dr Shami Chatterjee made arrangements to watch in case it returned. That's one of the risks in this kind of work, because it might have been a century before 121102 came back to life; in fact, however, there were nine flashes in six months (and there may have been more, because only 83 hours were devoted to observing the location of the pulses during those six months). Doctor Chatterjee was previously a Janskey Fellow at the Harvard-Smithsonian Centre for astrophysics and it was at the extremely powerful Carl G Janskey radio telescope array that the pulses were observed. Doctor Chatterjee's team was able to locate

121102 in a dwarf galaxy more than three billion light years distant from Earth.

Does this mean that the dwarf galaxy is home to intelligent life? It does not mean that at all (and nor does it mean the opposite, which would be that there is no intelligent life in that galaxy, or that these flashes were not sent by intelligent life. There is simply not enough information to say one way or the other). It is, though, helpful in ruling out some of the other possibilities that have been considered.

One widely held view of FRBs was that they were formed as a result of some cataclysmic event – the collapse of a neutron star into a black hole, for example, or perhaps a star exploding into a supernova. And it is possible that either of those events could produce the kind of burst seen from 121102 – but they could not do it repetitively.

It was also considered that FRBs were likely to be coming from within our own galaxy or, if not within, then very close by. And that may still be the case – for other FRBs – but at a distance of more than 3 billion light years from Earth, it is clearly not the case here.

(Our galaxy, which is known as the Milky Way Galaxy, has a diameter of between 100,000 and 180,000 light years).

It is possible that 121102 is a magnetar, a kind of neutron star that has a very powerful magnetic field. Neutron stars are small, very dense, and created by the collapse of a star with insufficient mass to produce a black hole.

But, when all of those caveats have been stated, it is also possible that some form of intelligence there is attempting to send a signal to intelligence elsewhere.

Records of **potentially habitable exoplanets** are kept, among other places, at the same Arecibo University that first caught sight of FRB 121102. They maintain a Habitable Exoplanet Catalogue (HEC). In addition, NASA has an Exoplanet Archive which stores, among other details, exoplanets found by SETI that are within their star's habitable zone and also possess other characteristics concerning physical phenomena, radiation, and star plasma.

Three things need to be said at this point:

- The fact that an exoplanet is in its star's habitable zone does not mean that it is habitable

- The fact that an exoplanet is in the habitable zone and is habitable does not mean that it is inhabited

- All the rules used by SETI and every other scientist involved in the search for extra-terrestrial intelligence are based on the idea that, to be able to support life, a planet must have a significant resemblance to Earth. The reason for this belief is that Earth is the only planet known to support life and that, clearly, is not an entirely satisfactory rationale for the belief. If an inhabited planet is ever discovered that is nothing like Earth, then the parameters will have changed dramatically and a much wider range of exoplanets will become candidates for habitation by intelligent life.

And two other things are also useful to remember:

- An habitable planet may host life

- Life may have been generated on an habitable planet.

Those are two different statements. The reason for making the distinction is that, until scientists have discovered more about life on another planet, the mere fact that it exists does not mean that life has ever been generated in more than one place. Scientists believe that life on Earth – the life of which humans are part – was generated on this planet. And that may be so. Or it may not. It is not beyond the bounds of possibility that the building blocks for life travelled to Earth from elsewhere. Perhaps on a meteor or asteroid; perhaps even on a long-ago spacecraft. It is also not beyond the bounds of possibility that the place they came from to reach here was not the first place in which their particular life form began to exist. The universe is 14 billion years old, give or take 21 million. That's more than long enough for life to have travelled – if not here, there and everywhere, then at least in a point-to-point journey taking in a number of stops on the way.

When all of that has been taken into account, a number of factors decide whether or not an exoplanet is regarded as habitable. The first and completely necessary factor is a source of energy. Life cannot exist without energy, which sustains metabolism. Beyond that, NASA says that the most significant criteria for an exoplanet to be habitable are: "extended regions of liquid water and conditions favourable for the assembly of complex organic molecules." And then there is the question of biosignatures, or "chemical fossils".

A **biosignature** may be a molecule, an element, or even something that a name cannot be put to in the state of knowledge we currently have, but that simply appears. A "phenomenon." If that sounds vague, it simply reflects reality. Almost everything that humans have a name for is something that, in the distant past or more recently, someone looked at and thought, "What *is* that? Where did it come from? What should we call it?" It would be easy to think that all of that naming of phenomena was done at the dawn of time, but the fact is that physicists today are seeing things never seen before and giving names to them. What makes a molecule, element or phenomenon a biosignature is that it contains evidence that can be scientifically validated that it either is now, or was at some past time, life. Evidence of life could be derived from the physical or the chemical structure, from the fact that it uses free energy, or the fact that it produces wastes and biomass. The point is that a biosignature is the product of living organisms. Once again, however, scientists must proceed with care because a biosignature might be universal but might also occur only on Earth. The fact remains, though, that there are chemicals (DNA is an example) that, if they are discovered anywhere in the universe, can only have come from some form of life.

Biosignatures are, by the nature of SETI's work, not part of its current search patterns. What SETI looks for is what it calls "habitability indicators":

- What the planet is made of – its bulk composition

- The planet's orbit. Some orbits are so elliptical that the
planet could be in the habitable zone for part of its
journey around its star, while at other times it could be so
close to the star that all possible lifeforms would be
incinerated and/or so far from the star that they would
all die as a result of the extreme cold

- The planet's atmosphere. Present assumptions are that an
atmosphere of pure hydrogen, for example, would be
inimical to life, though there may be a discovery just
around the corner that calls that into question. It's
certainly true that humans could not live in such an
atmosphere, but there is a danger of applying too
rigorously this idea that all life must resemble life on
earth. Given the ideas that humans have had in the past
about the nature of the universe, a little humility about
the state of human knowledge is recommended.

Those indicators are connected with the exoplanets. But what
of the stars around which the exoplanets rotate? There, too, SETI
has things it looks for to decide whether a star has a habitable zone
and whether life is likely to exist on an exoplanet in that zone:

- Mass. Is the mass of the star within the range thought
suitable?

- Luminosity. How much light does the start emit?

- Metallicity. "Metals" to an astronomer is not what "metals" are to you. The concern here is: what percentage of a star's mass is not either hydrogen or helium? The fact is that the great majority of matter in the universe is one of those two elements, and astronomers define "metal" as any element other than hydrogen or helium. A star that has significant volumes of carbon, nitrogen, and oxygen is said to be rich in metals; from the point of view of SETI, these are significant elements because they are regarded as essential in the formation of early life forms. Current theory says that immediately after the Big Bang, the early universe was almost entirely – or, indeed, entirely – made up of hydrogen and helium, with the other elements being formed within stars first by synthesis of hydrogen and helium, and then of the new elements that had already been synthesised.

- Stable variability. A variable star is one whose brightness (or apparent magnitude) is seen on earth to fluctuate. Extrinsic variability is less significant as a measure of the ability to support life on an exoplanet, because it is caused by something coming between the star and the observer on Earth. Intrinsic variability, on the other hand, is caused by swelling and shrinking of the star. If the variability is too great, conditions for life will not remain stable for long enough for life to form

PROJECT PHOENIX

The search for extra-terrestrial life began, in its present form, with NASA which, in 1988, started to fund the search for life in all areas of the sky. It was probably a coincidence that the actual search began in 1992, which was five hundred years after Christopher Columbus had landed in the New World. A year later, Congress cut off the funding. SETI was not prepared to give up and they looked for alternative sources of money to keep the search going. A great deal of the early funding came, and still comes, from donations from members of the public. There are more than 100 projects active at the present time; some of them are straightforward astronomical and planetary science projects, some are concerned with the evolution of the elements and other chemical substances, and some are research into climate change. There isn't any doubt, though, that the projects that have caught the public attention are those concerned with the search for life elsewhere in the universe.

When NASA withdrew its funding at the insistence of Congress, the search was named Project Phoenix, possibly in ironic reference to its rising from the ashes. In the beginning, Project Phoenix used the largest radio telescope antennae in the world to examine areas in space around a thousand stars closest to Earth with characteristics that suggested they resembled the Sun.

The SETI Institute entered into a joint project with the University of California at Berkeley to build 42 separate telescopes linked to perform as one single enormous telescope. It was named the Allen Telescope Array in honour of Paul Allen, Microsoft's co-founder, who funded it generously. The object was a detailed examination of up to a million stars over two decades.

Now, telescope arrays all over the world are engaged in the search, and SETI gets the word out quickly on any new possible candidate so that as many astronomers as possible will train their equipment on the source and, if anything is there to be discovered, discovered it will be.

WHAT SCIENTISTS THINK, AND WHAT JOE PUBLIC THINKS

For a scientist, discovery of even the simplest microbe on another planet would be indescribably exciting. It would not – necessarily – mean that the microbe was created on that planet; it might have been, or it might have travelled there on a meteor or a comet. It might even have been left behind by a visitor from another planet who arrived with a case of the sniffles. Be that as it may, a microbe is life, even if in a very simple form, and what scientists would now know would be that that exoplanet is capable of supporting life. There could very well be a Nobel prize for the person who found that microbe.

To a member of the public, though, the discovery – though no doubt important – is not the discovery she or he is looking for. After all – a microbe! "I had a bunch of those last winter when I had the flu. In fact, I have about 100 trillion of them in my own body and, by and large, they don't cause me any trouble. Go back and look again. And, this time, bring me what I want to hear about."

This neatly sums up the difference between two categories of extra-terrestrial life. On the one hand, any form of life, however simple. On the other, advanced civilisations.

Because microbes can't assemble transmitters and send signals into space, no telescope array, however large and however technically brilliant, can detect their existence. What SETI looks for is evidence that some other life form has evolved at least to the point that Earth has reached. Chapter 1 described the various shapes that such life forms could take and issued the caution that the "people" who build such a civilisation may look nothing like humans. May, indeed, have taken a shape that humans would find repugnant.

To a scientist, that is not the point.

There are a few other people for whom it would not be the point, either. There would have to be a re-examination of religious beliefs. That does not mean that religion would be disproved – it's possible to imagine a number of scenarios in which the evidence of life on an exoplanet and the form it takes makes religion stronger, although it would certainly also demand some changes in current doctrine. Humanists, too, might have to do some re-thinking. If it turned out that the most successful and developed civilisations are those that behave altruistically, that would send earthlings a certain message. If, on the other hand, success and development were built on constant warfare, the message would be very different.

WHAT CAN WE HOPE TO LEARN?

But how likely is it that the discovery that a civilisation existed on another planet would deliver that kind of message? If the planet in question was within Earth's solar system, it might just be close enough for us to learn more than the basic fact of the civilisation's existence. But that is not the case. There are three planets in our Sun's habitable zone: Earth, Mars, and Venus. Civilisation exists on Earth. It does not exist on either Mars or Venus. That isn't, really, any longer subject to dispute. HG Wells published The War of the Worlds in 1897. There were some fairly powerful telescopes on Earth at that time, but they were not powerful enough to settle the question. Astronomers now have much more powerful devices, and NASA has flown by Mars and Venus and studied both – and, in particular, Mars – in some detail. The idea that they may be supporting subterranean civilisations does not stand up and, if they were on the surface of the planet, they would by now have been seen and photographed. There will probably always be conspiracy theorists who believe that government is keeping from us facts that might make us nervous, but they don't deserve serious consideration.

So, if not Mars or Venus – where?

The answer is: too far away for any meaningful exchange of information. The nearest star to Earth is Alpha Centauri. It's 4.3 light years away. In astronomical terms, that's no distance at all – but if Alpha Centauri has a habitable zone, and if there is an exoplanet within that habitable zone that is itself habitable, and if an advanced civilisation has established itself there (and those are all big "ifs"), then it would take more than four years for a signal to travel from Earth to that planet and more than four years for an answer to come back. And that is to leave aside all questions of how Earth could phrase a statement or question in a way that someone in the orbit of Alpha Centauri could understand, and how they could phrase their reply so that we'd know what they were telling us.

But that assumes that what is possible for present-day earth civilisation is also what is possible for advanced civilisations elsewhere. And that might not be the case. Civilisations in Alpha Centauri's habitable zone may already have developed ways of communicating at speeds that people on earth can only dream of. They may also have developed the ability to travel at speeds unimaginable by us.

And those questions will take up a lot of the rest of this book.

Because, make no mistake about it: the search will continue. Nothing has been found yet but, as the SETI Institute itself has said, "The SETI Institute intends to press the search. Needless to say, the march of technology and new scientific discoveries will influence future SETI strategies. But giving up is not on the cards. Christopher Columbus did not turn around simply because he failed to find any new lands during his first few days at sea.

CHAPTER 3

THE KARDASHEV SCALE

Chapter 2 had some things to say about "civilisations" and "advanced civilisations". But what are those things? What do we mean by "an advanced civilisation"? The Kardashev Scale is there to answer that question.

Nikolai Kardashev is a Russian astrophysicist (he is still alive in 2017 at the age of 85). Born in Moscow in 1932, he graduated from Moscow State University at the age of twenty-two and began postgraduate studies in the University's Sternberg Astronomical Institute, completing his PhD in Physical and Mathematical Sciences in 1962.

In 1963, Kardashev took part in the first Russian search for extraterrestrial intelligence. While examining quasar CTA-102, he developed his ideas about what form extraterrestrial civilisations might take. They could, he realised, be ahead of anything on Earth by millions, and perhaps billions, of years. He developed the Kardashev Scale to define levels of possible civilisation. The original Kardashev scale had three levels (later researchers have added more).

The most important thing to understand about the Kardashev Scale is that it is based on energy. As Kardashev sees it, the level a civilisation has reached can be measured by the amount of energy it consumes.

In addition to energy, Kardashev focused on communications technology. (Those later researchers have also included other factors). In his paper, *Transmission of Information by Extraterrestrial Civilisation*, Kardashev said that an advanced civilisation would be able to transmit radio signals over great distances in space. The levels of civilisation he defined were:

- **Type 0 civilisations**. This is not, in fact, one of Kardashev's original three; it's added here because this is where Earth is at present.

- **Type I civilisations**, or planetary civilisations, are able to use all of the energy available on their own planet. This includes all energy the planet is capable of producing, and all energy (such as solar energy) reaching it from space. This is the level Earth has currently almost reached, though the present view is that a genuine Type I civilisation should also have developed the ability to control things like volcanoes, weather, and earthquakes and would also be likely to be building cities in the oceans. Clearly, then, Earth has not yet quite reached the level of a Type 1 civilisation; Carl Sagan put Earth's current level of development at seventy percent of Type I, so that Earth is, in reality, a **Type 0 civilisation**.

- **Type 2 civilisations** have reached the level at which they can make use of all of their star's energy, and not just the solar energy that reaches their planet from the star. Stars, of course, transmit energy in every direction and planets orbiting them receive only a small fraction of the total. How, instead, a planet might collect all of the energy emitted is an interesting challenge. In 1960, Freeman Dyson put forward the idea of the Dyson Sphere, a structure that would enclose a star completely, capture all of the energy it emits, and transfer it to the home planet. There's an interesting sidelight here: what would happen if two planets orbiting one star both had advanced civilisations, and one of them constructed a Dyson Sphere? That would dwarf the disputes over diverted water that we occasionally see on Earth. But the possibility of two advanced civilisations in one solar system is perhaps a little fanciful; a more interesting question is: what happens to the other planets when they

suddenly lose access to solar energy because it has all
been grabbed by the advanced civilisation in their midst?
Presumably, any early forms of life on those planets
would be wiped out. It's possible to imagine a Type 2
civilisation with ethical standards so high that they would
not feel able to do this. In fact, various Green political
parties here on Earth would be very likely to oppose it. A
cosmic rerun of construction delays while new homes are
found for colonies of the Great Crested Newt. (And, if
you think that's funny, just remember that the "people"
in a Type 2 civilisation may look a lot more like the Great
Crested Newt than they resemble you).

Clearly, a Dyson Sphere would be enormous – Freeman
Dyson himself suggested that it would cover an area 600
million times that of the Earth's surface. And the hope is
that that would make them easier to see, because one of
the things that searchers for extra-terrestrial life look for
is a Dyson Sphere in space. It isn't something that is
likely to have come about by accident. If we see one, we
can be pretty certain that we are looking at something
constructed by a very advanced alien civilisation.

How far are we earthlings from being able to become a
Type 2 civilisation, capable of building a Dyson Sphere?
Not, as these things go, very far at all; estimates are that
we may be able to construct such a thing, and thereby

make use of all the energy the Sun produces, between 1,000 and 2,000 years from now. 1,000 years ago, the Chinese were the first to use gunpowder in battle (and they had flamethrowers!). King Canute married his cousin Emma of Normandy, laying the seeds for the invasion of England by the Normans in 1066. Emperor Hadrian set up the first postal system and built a wall between England and Scotland. A thousand years isn't very long at all. You won't be here to see it, but it's still very imaginable. It's in the progression to a Type 3 civilisation that the number of years involved becomes monstrous.

- **Type 3 civilisations**, in fact, may be at least 100,000 years ahead of us and quite possibly more. A Type 1 civilisation is able to use all the energy its planet has. A Type 2 civilisation can use all the energy produced by its star. But a Type 3 civilisation can use all the energy produced by its own galaxy. And that is a huge step up. Every star in a galaxy is producing vast amounts of energy. A Type 3 civilisation must be able to capture and use all of it. And how many stars is that? It's a lot harder than you might imagine to count the number of stars in a galaxy. No one has ever yet found a way to count them individually, not least because no one has ever yet found a way to *see* them individually. If you were to Google the question, "How many stars are there in the Milky Way?", you'd get a variety of answers covering quite a range – it might be 100 billion, it might be 400 billion, it might be quite different from either of those numbers – because the fact is that no one knows for certain. What astronomers actually do is measure the mass of a galaxy (in itself a far from exact science) and then make some assumptions in order to deduce the number of stars represented by that mass (bearing in mind that the bulk of the mass will actually be dark matter – another mystery).

Something we *can* be sure of, though, is that there aren't any Type 3 civilisations in our galaxy. How can we be so certain? Because, if there were, they would have taken all the energy from every star in the Milky Way, and that includes our Sun. We would have no solar energy at all, and we'd all be dead. And what will happen to us if a civilisation elsewhere in our galaxy is so advanced that it is about to promote itself to Type 3? Then it's goodnight from all of us. But, if there were any chance of that happening, there would have been a Type 2 civilisation in the Milky Way Galaxy (our galaxy) for at least a thousand centuries and we would have seen its Dyson Sphere. So no need to worry for a while yet.

By the time our earthly civilisation gets around to becoming Type 3, it's likely that it won't be a human civilisation at all. Humans will have disappeared, replaced by some kind of mechanised being. There will be more to say about that before this chapter ends, but before we go there, let's take a look at what other theorists have suggested might be added to the Kardashev Scale.

Kardashev listed only the three types of civilisation described above. Others have proposed more.

- A **Type 4 civilisation** would harness the energy, not simply of a whole galaxy but of a whole universe.

- A **Type 5 civilisation** would find that very old hat and would be taking its power from a number of universes – a "multiverse."

- A **Type 6 civilisation** would have moved on to take control of space and time. Among other things, it would be capable of creating new universes. And that might be a good thing, because the chances are that by the time Earth reaches the exalted heights of a Type 6 civilisation, so many millennia will have passed that all the stars in this universe will have burned out and be cold and dead. Earthlings' descendants will need somewhere new to live.

As it happens, some scientists have decided to add a Type 7 civilisation to the Kardashev Scale. What capabilities would a Type 7 civilisation have? There's no point in even thinking about it.

Step back to that Type 3 civilisation, and these words:

BY THE TIME OUR EARTHLY CIVILISATION GETS AROUND TO BECOMING TYPE 3, IT'S LIKELY THAT IT WON'T BE A HUMAN CIVILISATION AT ALL.

What can that possibly mean? We'll be looking at this again in Chapter 9, but here are some of the ideas that are floating around.

THE DEVELOPMENT OF AI

Nearly seventy years ago, Alan Turing set out the Turing Test, in which a human and a machine would have a "natural language" conversation, watched by a human evaluator. The evaluator would know that one of the parties to the conversation was a machine, but not which one. The conversation would take place on a screen so that the evaluator had no hint from the sound of rendered speech. If the evaluator could not tell which party to the conversation was a machine and which was human, then the machine would have passed the Turing Test – it would be able to exhibit intelligent behaviour equal to or indistinguishable from a human's.

The evaluator would not have to decide whether any of the statements in the conversation were true (which might save the human party some embarrassment); only that the statement could have been made by a human as easily as it could have been made by a machine.

In a nutshell, that is a definition of AI (artificial intelligence). AI is still in its early stages of development on Earth. And yet, it's already present far more than most people probably realise. AI is not just about really good algorithms, of the sort that Google uses to decide which websites to direct you to when you ask it a question. AI – true AI – is capable of learning without any help. And we already have, here on Earth, examples of AI that can do just that.

Google has DeepMind. It extracts meanings and makes connections and doesn't rely on algorithms to do it. The more it's used, the smarter it gets. It *learns* as it goes along. Quantum computing (which operates in a way completely different from the binary computers we are used to) is at an early stage, but prototypes exist and quantum computing is likely to provide answers to some of the most difficult questions humans face. Whether humans will like those answers is another matter. Everyone with an Apple computer knows Siri, and the more you use Siri, the brighter she gets, because Siri, like DeepMind, learns on the job. Just like a human PA. Amazon has Alexa that is every bit as bright, though its true aim is – of course – to make Amazon even richer and more powerful than it is today.

And there lies the problem. It's possible to list a lot more examples of AI systems that, even today in their early stages, are capable of transforming the world and the lives of the humans and other creatures who inhabit it. The question is not, "Will humans build bigger and more capable artificial intelligence systems?" because the answer to that question is: Yes. They will. The question that needs to be asked is, "As artificial intelligence systems inevitably become bigger and more capable, will it be possible to control them?" Will they be used for the general good of all humanity? And, before you offer a positive and optimistic response to that question, here's something you might want to bear in mind. America will have advanced artificial intelligence systems. So will Russia. And Europe. And North Korea. And Saudi Arabia, and Iran.

Still feeling optimistic? Professor Steven Hawking's didn't. In an interview he gave to the BBC, he said, 'I think the development of full artificial intelligence could spell the end of the human race.'

He didn't mean, though, simply that competing ideologies would use AI to destroy each other, eminently possible though that is. What he said was, 'AI will take off on its own and redesign itself at an ever-increasing rate. Humans, who are limited by slow biological evolution, could not compete and would be superseded.'

The human brain is a quite remarkable thing. Computing speed is measured in floating point operations per second (flops), and an intelligent and quick human can actually reach 1 trillion flops, which sounds quite stunning – until you realise that the fastest supercomputers in 2017 execute almost 100 quadrillion flops. And they're getting faster all the time. The human brain isn't.

OPENWORM

The two most important things in the brain are neurons and synapses. A neuron is a cell and a synapse is a connection between neurons. The human brain is made up of about 100 billion neurons, connected by some 60 trillion synapses. The science of the brain is still at an early stage, but understanding is growing. OpenWorm is a project to map the synapses – the connections – in one of the smallest brains known, the brain of *Caenorhabditis elegans*. This roundworm is one millimetre long and its brain contains only 302 synapses.

When the researchers had mapped all of the connections, they made a little Lego robot. They put motors in it, and sensors for both sound and touch. And they also gave it a layout of neurons and synapses that precisely matched those of *Caenorhabditis elegans*.

Apart from the complex research into the connections (and imagine how much more complex will be the study of the much larger human brain), the most interesting aspect of this whole study was that the scientists did not give the robot any instructions. They didn't tell it to do anything. But it did plenty. And what it did was exactly what *Caenorhabditis elegans* would have been expected to do. Touch a food sensor, and it moved forward, just as the actual worm would have done, looking for food. Touch it on what it thought of as its nose and it stopped moving forward. Touch other sensors and the robot moved in various directions – once again, just as the worm would have done.

The conclusion would seem to be that what makes a worm a worm is the combination of neurons in its brain and the way they connect. Is there any reason to suppose that that does not also define what makes a human human?

And what happens if you take all those neurons and synapses from the human brain and – just as the OpenWorm project did with *Caenorhabditis elegans* – map them all into a robot? There are various theories about that. The Singularity is one of them.

THE SINGULARITY

The Singularity (or, to give it its full name, the technological singularity) is a theory that AI will, by triggering a growth in technology that will surpass anything yet seen, change human civilisation in ways that can hardly as yet be imagined. For example, a computer running AI will re-program itself and, having done so, will be likely to go on reprogramming itself in a series of cycles. The word was first used by John von Neumann who said, some sixty years ago, 'The accelerating progress of technology and changes in the mode of human life give the appearance of approaching some essential singularity in the history of the race beyond which human affairs, as we know them, could not continue.'

Science fiction writers can have a ball with this idea, but science fiction can become science fact. The idea that, once created, a superintelligence would keep upgrading itself is a possible step on the way to developing creatures that we may think of as post-human or may classify in some other way altogether and these could be the creatures that create Dyson Spheres around stars so far away that no human (as we currently understand that term) could possibly reach them. In other words, what has grown out of the human race but is capable of developing Type 2 and Type 2 civilisations would not look like us *or* like the Great Crested Newt. They might, for example, look like computers with sensors, hands, and some form of locomotion. Alternatively, they might look like something as yet unimaginable to the human mind.

And all of this has discussed only intelligence. But humans are more than intelligence. Feelings and emotions have not yet entered into this discussion. What about love? Hate? Altruism? Competitiveness? Those things and many more are all present in the human, and it has to be assumed that they, just like those unimaginably fast floating operations per second, are somewhere in those 100 billion human neurons and 60 trillion human synapses. It isn't very likely that those things will be left behind if (when) humans begin to transfer into other forms of life the essence of what makes them "them".

There isn't a great deal of comfort to be drawn from that. Human life, ever since it began, appears to have been marked by tribalism and the desire to conquer. Those things are so deeply embedded in human history that we have to assume that they are part of what makes humans human. Perhaps, after the singularity, as the superintelligences into which the human race has transformed itself develop ever greater levels of intelligence, they will see that the future lies in cooperation. In making sure that equality rules everywhere and that no-one is left out. But perhaps they won't. Perhaps what those superintelligences develop will be ever more efficient ways of conquering and destroying anything they see as a possible competitor.

Is that unduly pessimistic? Not if the history of the world is anything to go by. And it's time to be thinking about this, because the current estimate for the date of the Singularity is 2040. Which is not very far away at all.

Any reader who is now looking for reassurance may like to know that there are many researchers who believe that the Singularity will not happen. And there's something else to bear in mind, before the human race transfers everything it thinks of as human into machines. That something is solar flares.

SOLAR FLARES

Our sun is not a peaceful place. It is not a well-regulated furnace. The sun has "weather" and the weather can sometimes be stormy. It operates on a variety of cycles. Some of those cycles last seconds, some last billions of years, and most are in between those extremes. And we don't know just how long all of those cycles are.

One of the things that means is that we don't know how often very large solar flares are to be expected. But we know they happen. Solar flares are quite a regular occurrence. A flare is a sudden burst of radiation more powerful than the radiation the sun usually emits.

Something else that may happen in conjunction with a flare but may also happen on its own is a coronal mass ejection (CME). The corona is the sun's outer atmosphere and a CME is the sudden, violent release of up to 1 billion tonnes of matter at speeds that may approach many millions of miles per hour. They look as spectacular as they sound and any planetary body that gets in their way will feel the impact. And that also applies to a spacecraft. Which is one of the two problems that could have a serious impact on the growth to becoming a Type 2 civilisation.

Suppose that the robotic humans discussed above are out in space building Dyson Spheres when a CME catches them amidships. Every system on the spacecraft and in the robots will be fried. And now suppose that a flare much larger than Earth has seen during the past century occurs. Flares emit many kinds of radiation including x-rays and UV radiation. The largest there is any record for that involved Earth was in 1859, but it's certain there had been many before that date. The one in 1859 was called the Carrington Event in honour of the astronomer, Richard Carrington. According to reports from the time, the flare set fires on earth and melted the wires attached to telegraph poles. The effect of such a flare today would be far greater because it would render useless every integrated circuit on earth. Banking would be incapacitated. The control of weapons by every advanced army in the world, transport, navigational systems, the management of every organisation and company that relies on computers (and, today, which organisations and companies do not?) would be at an end. No-one would be paid, and the only way to buy anything would be by barter. There would be no air traffic control and anyone in the air at the time would be very lucky to make a safe landing.

In sum, rather than thinking about covering the last twenty-nine percent to finally become a Type 1 civilisation, the earth would be plunged back towards the middle ages.

1859 was a long time ago. Does that mean that flares of this sort are rare? It does not. They are happening all the time. Most of them leave the sun in a direction that does not bring them towards earth – but that isn't going to last forever. Earth had a near miss in

July 2012. There was a flare then that could have had all of the effects just described and it missed us, according to NASA, by not very much at all. Earth may be less lucky next time.

Flares of this sort are being sent out by stars all over the universe. It's impossible at this point to say how well future space voyagers from Type 2 or Type 3 civilisations may be able to protect their spacecraft against them. The assumption has to be that such protection will be possible, just as today's sea-going craft are not subject to the perils faced by the pirates of Byzantium. Because, otherwise, there won't be any Type 2 or Type 3 civilisations. (And perhaps there aren't. And perhaps solar flares are the reason there are none).

WHY IS THE KARDASHEV SCALE IMPORTANT?

The Kardashev Scale is important, not because it's an accurate description of three specific levels of civilisation that may exist in the universe, but because it provides a way of thinking about what SETI should be looking for. If you can't define something, you can't see it. In 2015, the Serbian astronomer M. M. Cirkovic published a study describing Kardashev's original work as, "A fine vehicle with room for improvement." He quotes Heraclitus as saying more than 2500 years ago, 'If you do not expect the unexpected, you will not find it; for it is hard to be sought out and difficult.' He also repeated an argument previously made by Claude Lévi Strauss to the effect that Darwin could not have done what he did if Linnaeus had not previously done what *he* did. Linnaeus, of course, laid down the form of taxonomy that is still in use. It didn't matter that Linnaeus had ideas about the origin of species that did not agree with Darwin's; what did matter was that Linnaeus provided Darwin with a framework inside which Darwin could organise his theories.

In the same way, Cirkovic argues that extra-terrestrial researchers need a framework to operate within and that, if they don't have one, they may very well miss an alien civilisation even when they are looking straight at it. He presents the following as the forms of search currently taking place that are all descended from the Kardashev Scale (he does not claim this as original; it is an amalgam of ideas agreed by a number of researchers in the field, and it's only fair to say that there are other researchers who disagree):

1. A search for signs of the technology that should be available to a civilisation more technologically advanced than anything on earth at this time. Such signs might be artefacts, manufactured products, a technical signature, or something else.

2. Signs of the development (see The Singularity above) of a civilisation into a post-biological and/or super-intelligent form. Such signs are more likely to be along

the lines of, for example, construction of Dyson Spheres than actually spotting a bionic person.

3. A willingness to expand the range of targets thought likely to reflect the existence of an extra-terrestrial civilisation.

4. A willingness to develop and expand the field of astrobiology and to establish greater contact and collaboration between astrobiology, evolutionary biology, computer science, and other related disciplines.

Something else Cirkovic says is that, "the energy values in Kardashev's scale should not be taken literally." It is convenient for people on earth to use the brightness of the sun, because that's what we are used to. In fact, astronomers use the brightness of the sun as the unit to measure the luminosity of other stars. The fact is, though, that most stars in the Milky Way Galaxy are less luminous than Earth's sun, and that is also true for most of the stars that have a habitable zone containing habitable exoplanets.

In other words, it's possible to take the Kardashev Scale too literally. And that is also true of a number of other things to be discussed in this book. Tribalism and a desire for dominance are not the only weaknesses on this planet. The people of Earth also have a tendency to arrogance – to the belief that, "This is the way we do things, so this is the way things should be done."

If the search for extra-terrestrial intelligence is ever successful, then denting that arrogance may be the greatest gift that humankind receives from the discovery.

CHAPTER 4

THE RARE EARTH

A lot of effort and a lot of money are going into the search for extra-terrestrial intelligence. What if it's all wasted? Suppose there *is* no one else? Anywhere, other than on Earth? We haven't yet found another civilisation. That could be because we haven't been looking for long enough. Or we haven't looked in the right places. Or we have, but we haven't looked for the right things in the right way – we haven't understood what another civilisation might look like. But it could also be that we haven't found another civilisation because there isn't one. Planets capable of housing complex life forms may be so rare that, for practical purposes, they don't really exist.

This is not a new idea. The title of this chapter comes from a book by Peter Ward and Donald E Brownlee who, in 2000, when they were both at the University of Washington, published a book with the title *Rare Earth: Why Complex Life is Uncommon in the Universe.* Their argument was that the evolution of multicellular life on this planet required a combination of circumstances so improbable that the chances are it has never happened anywhere else in the universe. The circumstances in question were geological and astrophysical.

Before going on with this, it's worth remembering that arguments are not always put forward for themselves alone. For example, Peter Ward has become an authority on climate change and, particularly, on the ability of humans to affect it. (These are two aspects of the same thing: that climate change is happening, just as it has ever since the earth came into existence, is undeniable. What is subject to disagreement is the extent to which humans (a) are responsible for the change and (b) have it in their power to control the change). Peter Ward uses the possibility that climate change will bring an end to life on Earth as a means to promote a particular set of views. He may be right. He may be wrong. But that is not the subject of this book.

FERMI'S PARADOX

The physicist Enrico Fermi gave his name to the Fermi paradox, although he was not the first to describe it. The paradox is this: estimates support the probability that civilisation exists outside this planet – so why has no one on Earth ever seen it? This galaxy alone contains billions of stars with similarities to our sun and some of them are far, far older (but see Chapter 8). Fermi and others forecast that large numbers of these stars would have habitable zones and that, in some of them, habitable planets would exist. As discussed in earlier chapters, those forecasts have now been shown to be correct. So where are the civilisations? Where are the aliens? Why haven't any of them turned up in the solar system? It's a question worth asking because, if our experience on Earth tells us nothing else, it tells us that intelligent life is (a) liable to find new habitats in which to live and (b) skilled at overcoming scarcity. There should be civilisations more advanced than Earth's that are already travelling in space and colonising new planets. And yet, none have as yet made themselves known. *The Rare Earth* asks whether that might not be because, however probable extraterrestrial civilisations might be, they don't actually exist.

It's possible to give the Fermi paradox more respect than it deserves. Yes, it is a question that has to be addressed – in fact, the purpose of this whole book is to address it – but there are those who say that, when converted from propositional logic to modal logic, the paradox disappears and that, in fact, it is not a paradox at all.

THE MEDIOCRITY PRINCIPLE

Although it didn't start there, the mediocrity principle is mostly used with reference to the way the solar system evolved, the history of Earth, and how biological complexity came to be and humans evolved. In other words, it has something to say about the human race's position in the universe. The word "mediocrity," though, has such a pejorative meaning that it might be better to rename the mediocrity principle as the normality principle and to say that, rather than a phenomenon being mediocre, it is merely normal. At the root of the principle is the belief that it is always better to assume that something is an example of normality and not that there is anything special or superior about it. The normality or mediocrity principle says that, since intelligent life exists on Earth, the assumption should be that it exists on Earth-like planets throughout the universe. That, in fact, the existence of intelligent life on planets capable of supporting it is the norm and not the exception. Renaming the principle does not remove the question: if intelligent life is the norm, why has it never been seen anywhere other than on Earth?

THE ACCUMULATION OF FACTORS OF CHANCE

The Rare Earth starts from the position that, for complex biological organisms to evolve, a large number of chance circumstances are necessary. Any one of them – any one, two, or three of them – might have happened again and again at various places in the universe. But what happened on Earth was that all of them came together, in the same place and at the same time. And the proposition in *The Rare Earth* is that the likelihood of all of those things happening at the same time in the same place is immensely unlikely.

To show the scale of coincidence and happenstance that led eventually to the evolution of humans on Earth, here is a list that is by no means all-inclusive. There is an argument here, though, that will be discussed again later but should at least be mentioned at this point, and it is: that this is what happened *on Earth*. The assumption most commonly made is that intelligent life could not also have evolved on Planet X unless Planet X underwent precisely the same sequence of events in the same order and within the same timescale – and that is an assumption that scientists on Earth are prone to make but for which no logical justification exists. It might be that intelligent life can only arise when all of these things happen, and happen in the same way. It might also be that there are countless sequences of events, some of which have been completely unknown on Earth and which, as yet, earthlings cannot imagine, but which also produce intelligent life. We don't know, and it is more of the human arrogance this book has previously mentioned to believe that we do.

That *caveat* having been mentioned, here is the list of what happened on Earth:

REVISITING THE HABITABLE ZONE: THE HABITABLE TRINITY

It's more than 50 years since the idea of the habitable zone was first put forward. Much more recently – In fact, 60 years later, in 2014 – James M. Dohm and Shigenori Maruyama of the Earth-Life Science

Institute in Tokyo have suggested that the original concept is insufficient. According to them, the concept of the habitable zone as normally defined does not provide sufficient criteria to decide whether or not a planet is habitable. They suggested replacing the habitable zone with the Habitable Trinity. This says that the minimum requirements for life to come into existence and to evolve are: **atmosphere**, **ocean** and **land mass** – and that all three are required. (This is why you will often hear references to "rocky planets with water" as places to look for life). In their words, "The presence of one component is not sufficient to create a habitable planet." They go on to suggest that scientists looking for planets outside our solar system that could possibly sustain life should regard the habitable trinity as a "baseline index." They point to the existence of technology that can characterise rocky exoplanets' surfaces through scattered light in order to obtain information on the nature of the surface including details of ocean, soil, snow, and vegetation.

At the root of this idea is the fact that life (that is, life as we know it, which is what Doctor Spock actually said, because he never at any time in any episode said, "It's life, Jim, but not as we know it") requires:

- Water (derived from the ocean)

- Carbon and nitrogen (derived from the atmosphere)

- Nutrients (derived both in the ocean and, probably more significantly, from the landmass).

They then go on to analyse what happens on Earth. At the equator, the ocean is warmed by the sun and evaporates to form water vapour which is moved by convection in all directions around the globe. Higher in the atmosphere, water is converted to ice and forms clouds, which later deliver rain; most of the rain falls in the foothills of mountain ranges because, as warm moist air climbs higher, it becomes cooler and cooler air holds less moisture than warm air. It follows that, if a planet is to give birth to life, it needs to have **mountains** and, if it is to have mountains, the planet also needs **tectonic plates**, because it is at the places where one plate grinds against another that mountains are thrown up. Weathering of the mountains, given the right kind of rock, produces nutrients that are washed down into floodplains making them fertile. Not every kind of rock will do; basalt is useful but is short of some of the nutrients necessary to life. Granite is a more certain source of nutrients, and Earth is rich in granite in high, mountainous areas where weathering is most efficient. Their data suggest that the nutrients provided by landmass, at least on Earth, are a million times more plentiful than those provided by mid-ocean ridges.

That leaves the question of **carbon** and **nitrogen**. Where do those come from? Today, the ocean is rich in compounds of carbon and nitrogen – but during the Cambrian Age, when life was developing on Earth, carbon and nitrogen in the early oceans would have been insufficiently soluble. The carbon and nitrogen that life needed came from the atmosphere. Carbon dioxide is now present in the atmosphere at only 400 parts per million (and that is a good thing, because human life could not cope with very high levels of carbon dioxide) but, in Earth's early days, it was much more abundant. And where the carbon dioxide has gone is important: it went into organic matter as the result of biological processes. Biological processes, of course, mean life, and so it becomes apparent that early forms of life themselves processed gases in the atmosphere in a way that made possible the development of more complex life forms.

BOTH THE STAR'S PLACE IN THE GALAXY AND THE PLANET'S PLACE IN THE STAR'S SYSTEM MUST BE IN HABITABLE ZONES

This is a lot less clear cut than it may sound, because it also concerns other planets in the same system. Life, if it appears at all, *and if it is correct to assume that what happens on Earth is what needs to happen elsewhere*, is going to arise on a relatively small, wet, rocky planet in the "inner" part of the system (that is, relatively close to the star) BUT also needs to have the right sort of planets outside it. For "right sort of planets," read "big gassy monsters." The sequence of planets in the solar system, starting closest to the Sun and moving outwards, is: Mercury; Venus, Earth, Mars, Jupiter, Saturn, Uranus, Neptune. The first four are all small, rocky planets, but Jupiter is not. Having Jupiter next after Earth is part of the secret of how life came to exist here. Jupiter is the largest planet in the Solar System. It's a massive gas planet and that great mass means that it exerts a very strong gravitational pull.

Before going further into this, step back a little. The discussion so far has been: How did life form? But there's a question that comes before that, and it is: How did planets form? The reason this question is important is that, so far as anyone can yet tell, life did not form in space. It formed on planets. And, as the discussion in this chapter shows, not just any planet. So, how did planets come to exist? And, because Earth is the only planet where life has as yet been shown to exist, how did Earth come to exist? And what – if any – was the influence of other bodies in the solar system?

A **planetesimal** is a word whose literal meaning is: "a small part of a planet." It is usually used to describe the small bodies that come into existence as part of a planet's formation, but it has another meaning which is: any small body (comets and asteroids are examples) left over from planetary formation. Current thinking is that planets form as the result of an accumulation of dust and small particles. One definition of planetesimal agreed on in 2006 by scientists studying the formation of planets (but not universally agreed by all astronomers) says that a planetesimal is a solid object that came into existence during the planetary accumulation process and has an internal strength dominated by its own gravity – that is, strong enough to hold it together – and the orbital dynamics of which are not affected by gas drag. Planetesimals in Earth's solar system are larger than one kilometre, because that is the size at which a body's gravitational influence can attract other bodies and the body can begin to grow. They contrast with protoplanets or embryo planets, which are large enough not only to hold themselves together by their own internal gravity but also to exercise a gravitational force on other bodies that come near them. Protoplanets generally have a size between 100 and 1,000 kilometres.

Ever since stars began to form, planetesimals have been crashing into each other and into other things. In Earth's solar system, the planetesimals doing the crashing have mostly been asteroids, and few people with sufficient interest in this subject to be reading this book will be unaware of the publicity given to asteroids that are scheduled to come so close to Earth that a collision is possible. The surface of Earth's moon and of planets like Mars show a history of asteroid collisions, and they have also occurred a number of times on Earth. It's possible (though it's a hypothesis – definite proof does not exist) that life began on Earth as the result of an asteroid carrying the building blocks of life from elsewhere crashing into the planet, so it is not beyond doubt that human civilisation may have developed on Earth as the result of an asteroid collision. That is as may be, and can certainly not be ruled out, but what is more certain is that life has been extinguished – at least once and perhaps more often – when a very large asteroid smashed into the Earth. The most commonly accepted hypothesis (though there are others) for the extinction of the dinosaur is that it came as the result of a huge explosion on Earth after collision with a very large asteroid.

Planetesimals are, therefore, a double-edged sword. It's conceivable that they created conditions on earth that made it possible for life to begin (and that is so even if the building blocks of life are not carried on the planetesimal). It is also undoubtedly true that too many collisions with planetesimals can destroy emerging life before it reaches a level of complexity that allows civilization to begin.

And that is where Jupiter comes in. Jupiter has been described as a sort of "celestial vacuum cleaner." Its huge mass gives it immense gravitational power, and the effect of having that powerful neighbour has meant that Earth has been spared a great many collisions with asteroids and other bodies, because before they could reach Earth they have been dragged into Jupiter's orbit.

And so it seems that the development of life to a complexity that can lead to civilisation requires a rocky planet close enough to its star to be in the star's habitable zone, far enough from its star for life not to be shrivelled up before it gets going, and also requires a massive gas planet further than it from the star but close enough to the planet to act as a janitor sweeping up the majority of planetesimals before they can get close enough to bring life on the planet to an end.

And how common is that in the universe? The fact is that it appears to be rare. As astronomers log more and more exoplanets in the universe, they find very few indeed that have that same combination of smaller rocky planets in the star's habitable zone PLUS a massive gas planet in the right position to carry out the janitorial function. As far as is presently known, only ten percent of all known stars have giant gas planets like Jupiter and very few indeed of those have the giant planet in the right position in relation to both star and small rocky planets.

There's more! It is not enough to have a massive gas planet like Jupiter in the right position in relation to a smaller rocky planet. The massive gas planet must also have the right kind of orbit. The orbit must be stable, and close to circular around the star. Of that small number – that fraction of the ten percent of giant gas planets that actually exist in relation to a small, Earth-like rocky planet – very few indeed have that sort of orbit. The conclusion is that a tiny fraction of the rocky planets of the right size in their star's habitable zone are protected by a planet like Jupiter in the right position and with the right orbit to protect them from having emerging life stamped out by a collision with an asteroid or other type of planetesimal.

There's more still! As if it were not enough that so few rocky planets are both in the right position and protected from planetesimal collision, there is a theory that simply having the massive gas planet in the right place and the right orbit is not enough. Konstantin Batygin is Caltech's Assistant Professor of Planetary Sciences. He and his colleagues have examined the series of events that they believe led to the development of life on Earth. Their conclusion is that more was required than simply mass, gravitational pull, position and orbit. They believe that the solar system as it stands now can only be explained if we accept that, at an early stage of its development, both Jupiter and Saturn drifted towards the Sun. If that happened, then large numbers of planetesimals would have driven planets larger than Earth towards the Sun and Jupiter and Saturn would also have dragged with them unimaginably large chunks of ice. Ice, of course, is frozen water, and that is where Batygin and his team believe the building blocks came from that later ended in the creation of Earth and the other rocky planets close to the Sun. Jupiter and Saturn subsequently drifted outwards again, taking up their present positions and giving Jupiter it's all-important circular orbit around the Sun in just the right place and with just the right power of attraction to draw planetesimals to it before they were able to collide with Earth.

And, as they say, once you accept that complex sequence of events, it becomes only too easy to believe that – immense though the universe is – it happened only once.

Note, though, that they do not put this forward as evidence that Earth is the only place that ever had the right combination of circumstances and the right sequence of events for life to begin. Serious astronomers do not possess that level of arrogance. What they say is that other planets like Earth are probably extremely rare anywhere in the cosmos.

Go back for a moment to the convection that moves the water vapour. It is caused by the heat of the sun and is a reminder that **energy** has not so far been mentioned as necessary for life – but necessary it most certainly is. Mars at one time had all the components that Earth had for life to develop – but Mars is much smaller than Earth and would have cooled much more rapidly. Its size also meant that it lacked the gravitational power to hold onto its atmosphere and its liquid water. Life may well have begun on Mars, just as it did on Earth – but it could not survive and evolve. Thus, the idea of a small rocky planet with water and the protection of a giant gas planet in the right place and orbit moves on to become:

A rocky planet **of the right size** *with water and the protection of a giant gas planet in the right place and with the right orbit*

Plate tectonics also require energy and it comes from the heat caused deep in the earth's core from radioactive decay.

Oh – and just how common in the universe are plate tectonics which, as has been shown, are necessary for life to begin? The answer is that we don't know but, as you might expect of the "Rare Earth," the answer is: probably not common at all. Evidence does not yet exist on the presence of plate tectonics in exoplanets – but we do know how many such planets exist within our solar system. There's only one. And we live on it.

RADIATION

It's possible to become far too concerned about radiation. Reading that sentence probably took you three seconds, and during those three seconds you were hit by about 45,000 radiated particles. This year, you will be hit by 500 billion. In a normal lifetime, that amounts to 40 trillion. And all of that comes from natural sources – natural radiation constitutes eighty-five percent of the radiation that most people experience. It hasn't harmed you. It's part of human life.

Nevertheless, radiation can be so dangerous that it wipes out life (think of the bombs dropped on Hiroshima and Nagasaki) and it can be so intense that it prevents life from forming in the first place. How much radiation a planet receives is an important factor in deciding whether or not life will begin and, if it begins, whether it will be sustained.

That means that the planets that are furthest from the Galactic Centre, or the furthest out in a spiral arm of a Galaxy, are the most likely to be able to give birth to and sustain life. This raises the question of the **Galactic Habitable Zone**.

A galaxy known as NGC 7331 is often described as a twin of the Milky Way. That may be true of a number of its attributes, but at the centre it has such high levels of radiation that complex life forms at the very least (and, quite possibly, simpler lifeforms) cannot survive there.

And so it becomes clear that the planets in a solar system subjected to high levels of radiation, however habitable they may appear at first glance, are not places where civilisation can be expected. What's more, even a solar system that is not subject to high levels of radiation has to remain that way for a very long time to allow complex life forms to develop. The search, then, should begin with planets in habitable zones around a star whose galactic orbit keeps it out of regions of high radiation. That probably means that SETI is not going to uncover civilised life in any system around a star that has an eccentric orbit – that is, one whose orbit is not as close as possible to a perfect circle. The star should also be one whose orbital velocity is close to or equal to the spiral arm's rotational velocity. Those who put forward the Rare Earth hypothesis to justify the argument that we are alone in the universe say that there is a narrow range of distances from the Galactic Centre at which that similarity of velocity can happen. Where those things are true, there is a good chance that the star and its solar system will either not drift into a highly irradiated zone at all or will do so only at mind-bendingly long intervals.

It won't come as a surprise after that introduction to the galactic habitable zone to learn that the orbit of the Earth's Sun around the centre of the Milky Way is an almost perfect circle, and that the period within which it completes its orbit is very close to the Milky Way's own rotational period.

Some scientists looking for extraterrestrial life put the number of stars in the Milky Way that occupy the galactic habitable zone as only twenty billion; others say the number is less than half that.

Once again – there's more! Of all the galaxies that have been seen so far, more than half have the same multiple arm structure as ours. Does that mean that more than half of observed galaxies are nurturing planets like Earth? No, it doesn't. There are strong arguments that say that this Galaxy – the one Earth lives in – is sufficiently unusual that at the very most only seven percent of other galaxies are likely to be like it.

Galaxies collide. Astronomers report frequent observations of colliding galaxies, although that is more a question of improved telescopes than of increasing numbers of collisions. Nevertheless, galactic collisions are a fact and it seems to be another fact that our galaxy has suffered far fewer collisions than most. That, in fact, ours is a peaceful galaxy compared with the majority. Some astronomers hold that the Sun is fortunate in never passing through a spiral arm, but at least one astronomer (Karen Masters, Associate Professor of Astrophysics at Haverford College in Pennsylvania) says that it does so roughly every 100 million years.

Something else that marks out the Sun from most other stars is that its luminosity varies by only 0.1 percent. It is also not a binary star, and binary stars are unlikely to support planets with sustainable life forms because the other star would disrupt its planets' orbits. At least half, and possibly more, of all stars are thought to be binary stars, which means that at least half of all stars have a vanishingly small likelihood of being orbited by planets capable of supporting life.

Small red dwarf stars can be written off as possible supporters of life, because any planets in their habitable zones will suffer something called "tidal lock." That means that one side of the planet is always facing towards its star, making it very hot, while the other is facing out into the black emptiness of space and will be very cold. Those planets are also at great risk of solar flares, which would ionise the atmosphere and have other effects very damaging to complex life forms.

After all that has been taken into account, it is still necessary to face up to the fact that only nine percent of hydrogen-burning stars in our Galaxy are "Goldilocks stars;" they are not too hot, and not too cold.

SO: ARE THE RARE EARTH PROPONENTS RIGHT? IS EARTH THE ONLY PLANET IN THE UNIVERSE WHERE CIVILISATION COULD HAVE DEVELOPED?

The simple answer to that question is the same as the answer we have given to a number of other questions: we simply don't know. In galactic terms, we haven't been looking for long enough. It may be that, after another couple of centuries or so, with telescopes and other technologies we cannot as yet even imagine, and 200 years of dedicated searching, it becomes necessary to say: We are alone. There is no one else anywhere in the entire universe.

Or it may be that we are on the verge of discovering multiple civilisations. We don't know yet. But we will. And bear in mind that the Rare Earth is only one of a number of theories, of which the most prominent will be the subject of the rest of this book.

CHAPTER 5

THE GAIAN BOTTLENECK

Chapter 4 talked about the Rare Earth theory, which says that Earth is the only planet where civilised life has developed because the combination of circumstances required for the development of life is so specialised – so rare in the coming together of all its essential parts – that such a combination has not been seen anywhere else in the universe. This chapter describes the Gaian Bottleneck, which takes a somewhat different view. It says that the combination of circumstances required for life to begin is not particularly rare and that life may well have started on many planets. Once started, though, it has not been able to continue. In essence, the Gaian Bottleneck hypothesis boils down to this: that the simplest forms of life have come into being in many places, but life is very fragile and the lifeforms that have existed other than on Earth have not gained control of their environment sufficiently rapidly. On earth, for reasons that the hypothesis attempts to explain (or, if not explain, then at least understand), life forms developed, became more complex, and survived. Everywhere else, they died.

We hear so much today about their potential to harm and even to wipe out life on Earth that it can come as a surprise to learn that greenhouse gases may well have been what made it possible for life on Earth to reach the stage it has. At first sight, it requires a leap of the imagination to understand how primitive life forms could have controlled their environment. A great deal of work has been done on this theory at the Australian National University, where Dr Aditya Chopra says that the environment on the great majority of planets early in their life cycle is unstable, and that life can only survive if temperatures at the planet's surface are kept stable. What kept them stable on our planet, and failed to do so on others, was water, carbon dioxide, and other greenhouse gases – but not on their own. They needed to be controlled by the earliest emerging forms of life.

That "control" has nothing to do with ideas of early man in rooms full of switches and dials; it's a matter of biologically regulating feedback cycles. There is an interaction between the earliest life forms and the atmosphere around them. Chapter 4 identified three small rocky planets in our solar system of the kind that might have supported life. Those three are Earth, Venus and Mars. In the distant past – some four billion years ago – it's possible that microbes, or at any rate single cell lifeforms which amounts to much the same thing, had begun to exist on all three planets. But then Venus became impossibly hot and Mars became impossibly cold, and that was that as far as life on those planets was concerned. Life on Venus and life on Mars was not able to stabilise the atmosphere on either of those planets – and yet, the theory put forward by the Planetary Science Institute at the Australian National University, and accepted as a working hypothesis by many other scientists, says that life on Earth succeeded in doing so. The rest of this chapter will be devoted to understanding why it was possible for life on Earth to achieve what life on Venus and life on Mars could not do.

And what is the Gaian Bottleneck? It describes the conditions that prevent life from getting a sufficient grip on its surroundings to prevent its early extinction.

THE EARLY AEONS IN A SMALL, WET, ROCKY PLANET'S LIFE

It will already be clear to anyone who has read this far that studies of cosmology and astronomy deal in unimaginably enormous periods of time. Gyr is a commonly used expression; it simply means 1 billion years. And it's only in billions of years that investigations of life on Earth, and, indeed, of Earth's history, begin to make sense. This table was produced by Doctor Chopra and his colleague, Doctor Charles Lineweaver:

<u>**STAGE IN PLANETS LIFE**</u>	<u>**CONDITIONS ON PLANET**</u>
First billion years	- It's hot, it's under constant and high bombardment from space, and nothing can live there
Next half billion to billion years	- It's cooling somewhat, the bombardment is reducing, and there is continuous loss of volatile substances
Next half billion to billion years	- Life begins to emerge, but the environment is not life-friendly, and so:
Next half billion to billion years	- Environment becomes so hostile that life is extinguished. Alternatively, and so rarely that it may have happened only once, Gaian regulation evolves fast enough for life to survive and carry out its own evolution. If this happens, life may exist for billions of years.

Life in those early millennia is competing in a rodeo. This is not, though, the Reno Rodeo, where contestants have grown up knowing how to ride a bull or keep their seat on a bronco. Single cell lifeforms come into existence and they don't have a clue what they are supposed to do. The planet goes through spells of immense heat, and then it freezes, and then it's hot again. The riders fail to stay in the saddle. They are thrown off. In short, they die.

As long ago as 1995, Nobel laureate Christian Duve in his book, *Vital Dust*, suggested that abiogenesis – the evolution from inorganic or inanimate substances of living organisms – may be what he called a "cosmic imperative." It's built into the universe; it has to happen. He pointed to the fact that water and energy are common in the universe. What doctors Chopra and Lineweaver suggest is that what decides whether intelligent life will evolve is not the difficulty in starting life, but the difficulty in keeping it going. Life begins. It evolves into a biosphere. But what it then needs to do is to evolve global mechanisms that will enable it to continue to exist and it needs to do so in a timescale that would seem glacially slow to us but, judged in cosmological terms, is rapid. And, almost invariably – in fact, so far as we know, with only one exception (Earth) – it fails. Life has come into existence, life has failed to sustain itself, life has departed.

Clearly, the Gaian Bottleneck is a different explanation for the lack of visible intelligent life elsewhere in the universe from those we have been dealing with up to now. In the earlier part of this book, we have discussed reasons why life may have failed to come into existence. Now, with the Gaian Bottleneck, we are examining the possibility that it may have come into existence often, but not been able to sustain itself. That the bottleneck is not with the emergence of life but with its continuance.

It is not, in other words, a matter of what the two doctors in Australia call, "the intricacies of the molecular recipe." It is that, when life emerges on a planet, its rate of evolution very rarely matches the speed required to regulate greenhouse gases and what scientists call "albedo" – that is, the amount of the total radiation from its star that a body receives that the body is then able to reflect. Radiation that is not reflected is absorbed, and absorbing too much radiation can be fatal. To quote once more from the two doctors, it would seem that the cosmic default for most lifeforms is extinction. It is not, as other hypotheses have suggested, a question of the degree of available light and the distance from a star; what makes maintenance of life on a planet possible over the longer term is how rapidly it evolves to reach a level at which it can maintain the habitability of the planet on which it finds itself.

It's a Catch 22 that may seem amusing so long as you are not one of the lifeforms on an affected planet: that a wet, rocky planet needs to be inhabited if it is to be habitable. Lifeforms, except in exceptionally rare cases, can only come into existence on a planet if lifeforms are already there. Another whole book could have been written if Yossarian had ever had to deal with that catch.

LIFE IS NOT A PASSENGER

The standard view in almost every other hypothesis is that life comes into existence as single cell lifeforms (microbes, if you like) and then develops over vast periods of time until intelligent life forms are shooting rockets and telescopes into space in the search for their interstellar fellows. These lifeforms are, essentially, passive – mere passengers on a vessel (their planet) that is racing through space and modifying itself or being modified by outside forces as it goes. The Gaian Bottleneck doesn't see things that way. In the Gaian view, what reduces the hostility of a planet to the continuance of life is the way that it itself modifies and regulates its environment. This would be the most extreme case of natural selection, with those lifeforms most capable of modifying and regulating their environment being the ones that developed. Feedback mechanisms that once existed on Earth and still exist in most parts of the universe are anti-life. They may not prevent light from coming into existence, but they snuff it out. What happened on Earth was that the lifeforms got their snuffing in first and dealt a killer blow to the forces that sought to eliminate them.

WHAT'S SO SPECIAL ABOUT OUR SUN? WHAT'S SO SPECIAL ABOUT US?

The answer to those questions may be: nothing. But they aren't questions that can be ignored. Why did it happen here if (as the Gaian hypothesis says) it happened nowhere else? The answer believers in Gaian regulation offer has to do with time. They talk about "unusually rapid biological evolution" and it's worth checking what "unusually rapid" means. Here on Earth in the human era, biological evolution in five generations (150 years for humans; a lot less for some small mammals but far, far longer for the cedars on North America's Pacific coast) would be considered rapid but, so long as the change was fairly minor, possible. When the Gaian hypothesis speaks of biological evolution as "unusually rapid," it's talking about the first billion years or so of the development of life.

So why? What was it about this start on this planet that caused those early single cell life forms to get control of their environment in a way that so many others on distant planets and in distant solar systems failed to do? And once again we come back to that seductive idea that the conditions we live with here on Earth are exceptional. Which means that earth and earthlings must be exceptional. Earth had sources of energy, it was bombarded from space less and less, it had the chemicals required to start life, it possessed a combination of circumstances that were especially favourable to the emergence of good-looking, intelligent people. People just like us.

And the Gaian hypothesis says that earthlings should stop patting themselves on the back, because their circumstances are nothing special. Countless planets had sources of energy that were just as good – and still have. Interstellar bombardment and bombardment from a host star began to fall for every planet in the universe; that's how the universe rolls. Earth's chemicals? They are scattered throughout space. And the combination of circumstances? Well – you may just be onto something there. But what you're onto is not an explanation of how life came to exist on Earth. It's more an explanation of why, on this planet and possibly on no other, life – having come to exist – continued, instead of being wiped out as is the predestined fate of life in this unfeeling universe.

HOW THE GAIAN BOTTLENECK HYPOTHESIS DEALS WITH OTHER THEORIES

In science, there are precious few – if any – facts. Einstein never said that the speed at which light travels is the maximum speed at which anything can travel, even though he is constantly quoted as having said that. All you will ever get out of a scientist is an hypothesis. A statement that, "This theory offers an explanation for all the facts we currently have on the subject. It stands up to all tests that we are at this time capable of putting it through. Until new facts emerge or new tests are devised, this is a working idea to explain what we presently know."

Those behind the Gaian Bottleneck believe that the bottleneck is a life continuation bottleneck. There are other theories.

The Emergence Bottleneck believes that life can only emerge when the ingredients for a complex recipe are all in place. They include chemicals, temperature, moisture, and a whole bunch of other things. And the recipe tends to be built around Earth. Obviously. This is the only planet on which life is as yet known to have emerged; therefore, the recipe for the emergence of life must be based on earthly conditions. Gaians look at that and say, 'Okay. So what is it that Earth has and other planets don't?' And the answer that emerges is, 'Not a lot. And quite possibly nothing at all.'

Start with the host star. Is the bottleneck to the development of life a *stellar bottleneck*? Is there something about Earth's Sun that isn't found anywhere else? Well, if there is, it hasn't yet been identified. Billions of other stars have the same sort of chemical make up as the sun. Vast numbers are orbited by small, wet, rocky planets of the same sort of size and at the same sort of distance from the host star as Earth. There are stars of similar size to the sun, with similar internal chemical processes and similar levels of flares and stellar storms. There's no sign at all of a stellar bottleneck.

Maybe it's an *element bottleneck*? Once again, there's no sign of it. Carbon, oxygen and nitrogen. To a lesser extent, but still important: sulphur and phosphorus. Those are the elements of life on Earth. They are also some of the universe's commonest elements. And they combine elsewhere, just as they do here. Combining two atoms of hydrogen with one of oxygen produces water. Hydrogen and oxygen are super-abundant throughout the universe, and they show no sign of being reluctant to combine elsewhere in exactly the same way as they do on Earth. There's no sign at all of an element bottleneck.

Detecting the necessary ingredients on remote planets is difficult, but it's unlikely that an *ingredient bottleneck* exists. The building blocks of life on earth are well known. Sugars, fatty acids and amino acids are all monomers, which means they form molecules that are capable of binding with other molecules. When amino acids bind, they form proteins. When sugars bind, they form carbohydrates, and fatty acids combine to form lipids. All of those products are essential for life. Nitrogenous bases are also monomers, but the bonding they do is a little different, because when a nitrogenous base binds with sugar and phosphate, it makes nucleotide monomers, and those are the building blocks for DNA. It follows that, if sugars, fatty acids, amino acids and nitrogenous bases are unique to Earth, then an ingredient bottleneck does indeed exist. To the extent that we can assume that what makes life on Earth would also be what made life somewhere else (and that's a big assumption), those building blocks must be present on other planets for life to emerge.

And they are! Or, it should be said, they almost certainly are. Those organic monomers fall to earth from space in meteorites. Not every meteorite – the meteorites in question are known as carbonaceous chondrites, which are some of the most primitive (or earliest) meteors known to exist. They fell from the sky in the earliest of times, and they are still falling now. And not only on Earth – this planet and this solar system do not have a monopoly on carbonaceous chondrites.

These meteorites were at their most common and falling most often in the first billion years of Earth's formation. There is no reason at all to suppose that they did not and do not fall in the same way on other planets in that early stage of their existence. What they bring with them is what they brought to Earth. We might call them prebiotic molecules – those same building blocks as described earlier in this section that are the precursors for biological molecules.

As far as we can tell, then, there is no sign of an ingredient bottleneck.

How about energy – is there a *free energy* bottleneck? It seems unlikely. It isn't true, as was once thought, that all living things derive energy from the sun, because living organisms have been found in the deepest places in the oceans growing and surviving from energy emerging from vents in the seafloor and coming from deep within the earth. Nevertheless, energy is vital to life, and it has to come from somewhere. At first sight, it seems unlikely that there is much energy to be had from starlight inside a black hole, although that is not as obvious as might appear because black holes are packed full of energy from all the stars, planets and other bodies that have been drawn into them. But all the stars that we can see are emitting energy, both as light and in other forms, just as the sun, earth's host star, does. And research makes it clear that everywhere in the universe there are rocky planets with dense and very hot interiors creating thermal gradients driving energy to the surface. It would appear that the idea of a free energy bottleneck is a nonstarter.

That leaves us with a **Recipe Bottleneck**. Is life rare in the universe because there is a "recipe" that lists all the things that have to be in place before life can start to exist, and because all the ingredients for the "recipe" come together incredibly rarely? And here we run into something of a difficulty, because it isn't possible to know exactly what kind of chemistry with what kind of raw materials in what kind of environment led to the creation of life on Earth.

And if we did know the answers to those questions, we'd run into a second difficulty, the Dracula Difficulty, which is: How is it possible that so many scientists with clear theories about the raw materials, the environment and the processes needed to create life have not, in fact, been able to bring into existence even the simplest single-cell amoeba in the laboratory? One answer to that question would be that there is indeed a recipe bottleneck in the way of the creation of life, and it lies in the fact that the recipe is so complicated that not only can Earth's scientists not put it into practice but whatever the equivalents of Mother Nature may be on other planets have not been able to do it, either. Some would say that that explanation is a little fanciful.

de Duve, who has already been mentioned in this chapter for his description of life as a cosmic imperative, said that life was either a reproducible, almost commonplace manifestation of matter when the right circumstances and conditions applied, or it was a miracle. He didn't believe that anything between those two extremes was possible. He didn't believe in miracles, either. (That doesn't mean the rest of us can't. If we don't find an answer to this question, then the idea of a miracle just may be the elephant in the room. Ideas about divine creation are rejected by what is probably the majority of educated opinion because, "we now know too much about the science of life." We do? It isn't always obvious).

Is it necessary to conclude that the idea of an emergence bottleneck has to be abandoned? That is certainly the position of believers in the Gaian Bottleneck. And, if the arguments for an emergence bottleneck are as weak as the discussion above makes them appear, then it is necessary to find another reason for the failure so far to find signs of life elsewhere in the universe.

The Gaian Bottleneck is such a reason. It accepts that there are no emergence bottlenecks and, indeed, that emergence is common. Emergence – the formation of life – is, in the Gaian argument, probably very common all around the universe. Life begins. And then it ends. Without there ever being sufficient development for the emergence of an alien Socrates, Shakespeare, or Wayne Gretzky.

THE THREATS TO THE CONTINUANCE OF LIFE

It's worth bearing in mind that we don't actually know for certain
that the first life forms on earth did survive. Between 4½ and 4
billion years ago, Earth was a rather different place from what it is
now. The planet went through spells of almost unbelievable cold,
interrupted by spells of unimaginable heat – so hot that Earth was
covered by oceans of magma. One of the reasons the temperatures
were so unstable was that the planet was subject to frequent impacts
from space. And so, there were times when Earth was habitable, and
times when it was not. It's entirely possible that primitive lifeforms
emerged in those early habitable periods only to be bombarded,
frozen, and boiled into extinction. There's no way of knowing.

The Gaian Bottleneck hypothesises that the same thing
happened almost everywhere in the universe – but that, probably
uniquely, here on Earth a habitable period appeared and continued
for long enough in conjunction with habitable conditions (which are
by no means the same thing) for life not only to come into being but
also to develop to the point where it managed its own environment
and secured its own survival.

How did that environmental management come to be? It isn't
really possible to say. Something we do know about how nature
works is that there appears to be in living organisms an inbuilt urge

to continue by passing their DNA to the next generation. Enzymes came into being and evolved in a way that allowed them to manage the conditions they found themselves in. It sounds fanciful – but it's no different from the way camels developed the ability to store water through periods of drought that would kill most animals, or the way that plants evolved to secure their own future. How many people know that the grasses that are so ubiquitous in today's world are the latest stage in an evolution that began with lilies? A lily is a beautiful plant, but it relies for the survival of its species on the arrival of insects at just the right time to transfer pollen from stamen to pistil (and, in the case of some species, the formation of little bulb offsets around the base and the breaking away of scales from the bulb). By the time the line had extended to grasses, all that had been done away with – the wind does all the pollination that's needed.

During that period between four and half billion and 4 billion years ago, the size, weight and frequency of the bombardment of Earth decreased by a huge amount – some scientists say that it was by more than thirteen orders of magnitude. Conditions became more favourable to the emergence of life. Did this happen only on Earth? Only in our solar system? It seems extremely unlikely. Rocky, wet planets everywhere in the universe must have experienced a point at which they were cooling and they had environments in which life could emerge. Re-emerge, perhaps, because it's very likely that some of the earliest life forms began deep in the ocean or in vents in the earth, to which they would have been driven by the intolerable bombardment of the surface.

What the Gaian hypothesis says is that there is a natural tendency in planets to move away from habitability. They lose their volatile components, and that means they lose water and they lose atmosphere. It isn't easy for a planet to retain water on its surface. A greenhouse effect can see surface water evaporating from the surface into the atmosphere and from the atmosphere into space. The opposite – a deep freeze – turns water into a form in which it is not available for forms of life to drink it. This is probably something that happens everywhere; water vapour is being lost right now from Earth's upper atmosphere as a result of ultraviolet radiation and eventually all the water on earth will be gone and life will be gone with it.

But the accepted timescale for that to happen is between one and two billion years from now, and life – human life – will either have run its course or will have colonised other planets. That makes Earth the exception. Venus and Mars almost certainly both started out in the same way as Earth did, but Earth held onto enough of its water to sustain life over the long term, while Venus lost almost all of the water it had. Mars lost slightly less – only about eighty-five percent – but the rest froze into the polar ice caps and a permafrost below the surface of the planet.

The question is: if that happened to Venus and it happened to Mars, why didn't it happen to Earth? The Goldilocks principle has often been offered as an answer: Earth was just the right distance from the sun, just the right size, had just the right gravity, and was subject to just the right amount of light and heat radiation. The Gaian Bottleneck hypothesis doesn't accept that; what it says is that life got a start on Earth and developed quickly enough to gain a sufficient degree of control over the environment to keep going.

EXTRATERRESTRIAL RESEARCHERS SHOULD LOOK FOR WATER ON THE SURFACE OF OLD PLANETS

The Gaian Bottleneck hypothesis is not widely accepted by large numbers of specialists in the search for extraterrestrial life, but it does have a following and it does have a logic of its own. The word "biosignature" has been coined to mean something that might indicate the presence of life – that life is there now or has been there in the past. Carbon and oxygen have both been mentioned as possible biosignatures, because both are necessary for life – but it's possible that insufficient attention has been paid to water – liquid water on a planet's surface – as a biosignature.

If the Gaian hypothesis is correct, then there will be chemical equilibrium in the atmospheres of most or all old planets. By equilibrium, we mean that nothing is changing; the atmosphere is balanced and looks as though it has been that way for some time and will continue in that state for some time more. If the atmosphere is in equilibrium and the planet is not a young planet, then the Gaian hypothesis says that the planet is uninhabited. If we find an exoplanet with surface temperatures capable of supporting water, it will be worth examining the atmosphere for signs of disequilibrium. If we find an exoplanet with surface temperatures capable of supporting water and with an atmosphere in disequilibrium, it will be worth looking very closely for signs of life, because it is life that controls the environment to keep the water there, and it is life that upsets the equilibrium of mature atmosphere.

If, that is, the Gaian Bottleneck hypothesis is correct.

CHAPTER 6

THE GREAT FILTER

Robin Hanson is an economist. He is also a serious thinker on many subjects outside economics, including the possibility of life existing elsewhere in the universe. The Great Filter, which has become a staple of thinking about extraterrestrial life, was originally the name of a paper written by Hanson.

Life develops from abiotic (that is, non-biological) matter, often called "dead matter" though that can be confusing because "dead" suggests something that was once alive and abiotic matter has never lived. It contains some of the raw ingredients for life and, if the rest of the necessary components come together with abiotic matter, then life may possibly begin. If it does, it will be in its simplest form – the one-cell lifeform – and, if it survives long enough, then in four or five billion years it may have developed into multi-cellular organisms sufficiently complex to be capable of forming a civilisation.

But that clearly doesn't happen very often, because if it did we'd be able to see evidence of such civilisations elsewhere in the universe and, so far, they have eluded us – possibly because they aren't there. This book presents many theories as to why the universe should be so empty of advanced civilisations, or even simple lifeforms, and the Great Filter is Robin Hanson's contribution to the debate. Hanson describes the idea of "a great filter between death and expanding, lasting life," and poses a question that he describes as ominous (and you may agree): How far along this filter are we?

As an economist, Hanson is known for casting a sceptical eye on generally accepted theories. He has done the same with theories on extraterrestrial life. He says that physicists, astronomers, biologists and social scientists have all advanced theories that would suggest that the filter should be much shorter than it is. Hanson is not one to attempt to modify the results of the experiment to fit the theory; what he says is that, since the reality is a filter much longer than the theories predict, one (at least) of the theories must be wrong. His conclusions are not comforting – he says that humanity must be "much more wary of possible disasters."

COLONISING APPEARS TO BE CENTRAL TO LIFE

While, as before, bearing in mind that there is a certain arrogance in assuming that what happens on Earth must be what happens everywhere, it is not possible to ignore what simple observation on this planet can tell us about the nature of life. And the fact is that life on earth has adapted and evolved in ways that allow it to fill every ecological niche there is. Insects, plants, fish, people and bacteria – they don't stand still. They don't reach a stable population level and stay there. There is what must be assumed to be an inbuilt drive to expand. Lifeforms both simple and extremely complex have made astounding adaptations to expand into new ecological territory. (Sometimes, they do so by collaborating with the population already there, but at least as often they expand by conflict, and there's a lesson there, too, from which the human race needs to learn. Conflict is not an aberration; it is a norm. As long as there are other tribes on the planet, you disarm at your peril).

The relevance of this is that, elsewhere in this book, we discuss theories that say there are alien beings out there, and we just haven't seen them yet. That idea can be sustained, as long as it's accepted that the alien civilisations have not yet reached very advanced stages – say, Stage 2 or 3 on the Kardashev Scale. Because, if they have, and if there's anything in the assumption that what happens elsewhere is probably very like what happens here, then they should be colonising. And we should be able to see those colonies, even if we have not yet become one ourselves.

Also affected is another theory found in this book, which is the one that says that there are advanced alien civilisations and we haven't seen them because we have nothing to interest them and they can't be bothered to get in touch. Perhaps these are aliens who don't go to dinner parties. Or they do, but they don't consider us, the people of Earth, as the kind of people they want to spend an evening with. And, once again, it doesn't tie in with the nature of humanity. If extraterrestrials take that line – if they simply don't think we are interesting enough for them to spend time on or with – that shoots down in flames the assumption that what happens on Earth is very similar to what happens elsewhere. Because a huge amount of human colonisation has gone on on Earth over the centuries, and it had nothing to do with a belief that the people being colonised had something to offer (at least, apart from minerals, plants and other forms of raw material). When the Dutch colonised what is now Indonesia, they didn't do it because they wanted to spend quality time with the local people. The British colonised India and attempted to colonise what was then Mesopotamia for reasons that had everything to do with profit, but profit was not involved when they raised the flag over Tonga or Christmas Island. And the movement to North America of what were then English and Scottish people had nothing to do with profit; it was entirely about finding a new place to live. Oh, sure – their governments followed them wanting to hand out charters in exchange for money and demanding taxes, but what they got in return for that was the American War of Independence. New Zealand was settled before the British came by the Maori, who got there over huge distances of ocean in flimsy vessels – risky journeys they didn't *need* to make and at the end of which they were no better off in material terms than they had been before they set out. We could go on. The point is that people on earth move. They migrate. They go elsewhere, not just for money, and not just because things are not too good where they are, but because migration is a deep-seated human instinct.

Nor is it just a human instinct. Because so do animals, and – admittedly rather more slowly – so do plants.

So, if migration is a fundamental part of life on Earth, and if the assumption is that the drivers of life on Earth would also be the drivers of life on far distant planets, why have we seen no evidence of colonisation and migration anywhere in the universe?

AND SO IS KILLING

Murder goes all the way back to the emergence of the first living creatures. Cain and Abel were the sons of Adam and Eve, and Cain murdered Abel in jealousy that God had preferred Abel's sacrifice over his. One of the sons of the very first human beings on Earth murdered the other, and you can't go much further back in the human story than that!

All right, so Adam and Eve and their two boys are a Creation myth, told by early people to explain how mankind came to have populated the planet. But think about those migrants to Indonesia, India, New Zealand, and North America. They weren't the first inhabitants of those lands. When the Maori landed on the shores of New Zealand, they couldn't take possession until they had either killed the people who were already there or mated with them. (In fact, they did both). European settlers in North and South America were responsible for catastrophic collapses in the number of people they found there. Some of those deaths were because the Europeans brought with them diseases to which they had built up immunity, while the local people had none – but disease was by no means responsible for all the deaths. The settlers themselves killed an awful lot of people. But those people who died can only be called truly indigenous if their ancestors were the humans who arrived in the Americas between 50,000 and 17,000 years earlier during the Wisconsin Glaciation, when the sea fell far enough and the ice built bridges strong enough to allow them to cross. And they weren't. The ancestors of the people who died had wiped out the people they found there. If we know anything at all about the human race, it is that it has what seems to be an inbuilt tendency either to enslave or to kill other humans from different tribes.

And the killing was going on long before humans appeared on Earth. Some of the earliest animals were herbivores but a great many ate meat, and they didn't buy it in the supermarket, and nor did they restrict themselves to only eating the flesh of animals they came upon after they had died. They butchered their own dinners. And you can go back further than that, because some of those early single-cell organisms were bacteria and they got by by invading and killing other single-cell organisms.

WHY HAS NO ONE ENSLAVED OR KILLED THE PEOPLE OF EARTH?

The reason for the last section was not to spread fear or to encourage bans on extraterrestrial exploration for fear of what the civilisations we find might do. (If they have the technology to colonise this solar system, they already know where we are). It was to say that the fact that Earth has not yet been colonised and its people enslaved or wiped out is a reason to believe that there is no civilisation out there capable of doing it. Because, unpalatable as it may be to realise this, if the extraterrestrials are anything like humans, that is exactly what they *would* have done. They would have come here, "because it's there," because that's what the people of Earth do, and they would have set about enslaving and exterminating the people they found here – because that's what the people of Earth do.

It is, of course, possible that an advanced extraterrestrial civilisation does not take the same warlike stance as humans. It's also possible that an extraterrestrial civilisation, when it was less advanced than it is now, did take the same warlike stance as humans but has now developed beyond that – possibly after a calamitous outbreak of fighting on its home planet. (For an example of how that can work, look at France. The religious wars in France between Catholics and Protestants were so vicious that in 1572 in the Saint Bartholomew's Day Massacre, egged on by the king's mother who was a Medici, Catholic mobs murdered anywhere between ten thousand and seventy thousand French protestants. If this sounds like small beer in today's terms, we should remember that (a) the population of France at the time was only about fifteen million and (b) this slaughter came after several others almost equal in size in what had been a bloody hunting down of anyone who refused to accept the state religion).

And, although it didn't happen immediately, that was the reason for the rigid separation between state and church that still exists in France today. Sometimes it can take the bloodiest outbreaks of violence for people to realise that the cause of the violence should be put aside. The French succeeded in removing religion as a cause of violence, and it may be that an advanced alien civilisation has succeeded in removing tribal difference (in this case, the difference between them and us) as a reason to wipe another civilisation out. Before we become too comforted by that, though, we should bear in mind the waste laid to large parts of Europe by the French army in the time of Napoleon. It wasn't ideas about the delights of peace that brought Napoleon's mayhem to an end; it was the warlike Duke of

Wellington supported by the British Army and the Prussians – two more groups not famed for abhorring violence.

The murder rate in England today is roughly 1 per 100,000 inhabitants. In Queen Elizabeth I's time in the 16th Century, it was 5. Three hundred years before that, it was 20. So it could be argued that at least some parts of human civilisation have become more peaceful and less murderous over the centuries. It could also be argued that the urge to colonise and dominate is as powerful as ever, and that people have developed new ways of dominating and enslaving others without needing to get blood on their hands.

WE WOULD COLONISE – WHY HAVEN'T THEY?

Enough talk of violence and murder. Humans are already exploring space, but warm human bodies are only present on the shortest voyages to the Moon and into orbit around the Earth. For now. The merest glance at the human race's history and habits says that, as soon as it has reached the requisite point on the Kardashev Scale and is capable of space travel to more distant parts of space, people will be going there. The space travel that is currently possible is not restricted to government. Billionaires offer other very wealthy people the opportunity to travel briefly into space – if not right now, then very soon. We can be fairly sure that that will continue. When a future Elon Musk or Richard Branson asks people if they're interested in a journey to Mars, and a Richard Branson or Elon Musk even further into the future advertises journeys (they'll almost certainly be one-way) to planets beyond this solar system, any survey of earlier human behaviour says that some people will buy tickets. Not everyone – however bad things may sometimes have been in Ireland, most people stayed there instead of getting on a boat to Liverpool or New York – but some people did emigrate from Ireland, and some people will emigrate from Earth. They'll do it because of that innate drive to see what it's like somewhere else.

All of which tells us that a sufficiently advanced civilisation elsewhere in this galaxy or – if it sufficiently advanced on the Kardashev Scale – in another galaxy should be migrating to other planets, one of which (since this has proved itself to be a habitable planet) should be Earth. And they haven't. And Robin Hanson, when he talks about the Great Filter, invites us to consider some of the less attractive reasons why they may have failed to do so.

THE GREAT FILTER

Robin Hanson lists the various steps of biological development from the beginning to where we are now. He starts with the right star system. What's needed next is something like RNA, the ribonucleic acid without which, for example, genetic information can't be encoded and passed on. Single-cell life begins and is followed by lifeforms that are still single-cell, but more complex. Then we have sexual reproduction which is necessary before multi-cell life can begin to exist and, after immense periods of time, animals with larger brains begin to use tools, and that is the emergence of technology. And then, Hansen says, we get to where we are now (he does have the grace to admit that this is not an exclusive list of the stages the Earth and its humans have gone through). There will be a next step which will be a combination of technological and biological development leading to what he calls a "great explosion" which will resemble the great explosion that Columbus began, though this time it will be into space and not across the oceans.

All of these things have already happened here on Earth. Except the great explosion that takes us into space. Hanson points out that forming and then verifying hypotheses for all of the biological steps to date has employed some of the finest scientific minds, and that the explanations that have been put forward for these steps are entirely plausible. He also says that, somewhere along the line, "Someone's story is wrong."

Our universe seems to be almost entirely dead. The conclusion must be that immense difficulties stand in the way of the development of life forms so advanced that they become "explosive." (A better expression for explosive in today's world might be "viral," with its social media suggestion that life finds a pathway that is simply irresistible and is "shared" again and again until it fills the universe).

Hanson takes his hat off to the biologists and other scientists who have developed such utterly believable theories for how life develops. (For an example of just how plausible such theories can be, take a look at Richard Dawkins's explanation of how the seeing eyeball came into existence. There could be no more rational illumination (excuse the pun). And it has the twin advantages that it cannot be tested and cannot be disproved. But is it right?) And, as Hanson points out, the plausibility of the theories has launched as many estimates of the countless numbers of advanced civilisations that must be located somewhere in the universe as the beauty of Helen of Troy launched ships. The ships found their target. The searches for advanced civilisations so far have not.

IS THE THEORETICAL ERROR BEHIND US? OR AHEAD?

It is difficult to argue with Hansen's view that the fact that the universe appears to be almost entirely dead (the exception being here on Earth) means that at least one of the steps in the "filter" – that is, the steps that lead from finding the right star system all the way through to the beginning of an explosion into the colonisation of space – must be a lot less easy than theorists have so far said it is. But which? If a step that Earth has already passed is much more difficult to achieve than has been supposed, then life forms that have developed independently are that much less likely to exist.

But suppose the Gaian Bottleneck hypothesis is, at least in part, correct, and that the development of life to a certain point is not only easy but inevitable. That will mean that the block – the critical event in the filter – is ahead of us and not in our past. It's possible that we may never develop to that Stage 2 on the Kardashev Scale that will see us exploding into space and colonising distant solar systems, because something is going to intervene to prevent that development.

Hanson recommends close examination of the theory behind evolutionary steps that Earth is believed already to have passed to see whether any of them look more plausible than, in fact, they are. He thinks finding such a stage would be positive, because if it turns out that all previous evolutionary steps are as simple as current theory has suggested they are, then the inescapable conclusion must be that the filter is longer than it seems and that hurdles lie in front of us that are sufficiently challenging that no previous civilisation anywhere in the universe has been able to overcome them. And if that is the case, then the future for humankind is bleak.

But perhaps (and, unlike most economists, Hanson is nothing if not an optimist) it will be possible to make our prospects a little less bleak. Hanson recommends close examination of some of the things that can go wrong. Nuclear war. Ecological disaster. Failure in the security systems surrounding nuclear energy generating plants. The escape into the wild of a toxin developed for military purposes. A fatally mutated virus. There is a very long list of disasters that could befall this planet and wipe out all the life on it, leaving Earth as dead and barren as everywhere else in the universe appears to be.

THE DRAKE EQUATION

The Drake Equation was written as a way of estimating how many civilisations, advanced to the point where they are capable of communicating, might exist in the universe:

$$N = R^* \bullet fp \bullet ne \bullet fl \bullet fi \bullet fc \bullet L$$

Where:

N =   The number of civilisations in our galaxy issuing detectable
electromagnetic emissions

R* =   The rate at which stars suitable to host planetary systems on
       which intelligent life could develop are forming

fp =   The percentage of such stars that have planetary systems

ne =   The average number of planets in the solar system that have
       an environment in which life could possibly develop

fl =   The percentage of such planets where life does, in fact, occur

fi =   The percentage of planets on which simple life forms exist
       that then go on to nurture intelligent life

fc =   The percentage of civilisations that develop technology
       capable of emitting into space detectable signs that they exist

L =   The period of time during which such civilisations are
       releasing into space detectable signs that they exist

It's always good to have an equation. This one is popular among
seekers for extraterrestrial life. It has an impressive number of
variables, all of which can be justified. In fact, it suffers from only
two drawbacks: the fact that it is not possible to put a sensible value
on a single one of those seven variables; and the fact that a number

of the variables have more to do with biology than with astronomy –
and Frank Drake, who wrote the Drake Equation, was an
astronomer.

Biologists have been challenging the equation since it was first
published in 1961. One of the variables they challenge, though it is by
no means the only one, is fc, which has to do with the development
of technology. All technology on Earth, including the most advanced,
began when early humans first began to use the simplest tools. We'll
never know what the first tool was – perhaps it was a sharp flint
capable of cracking open the shell of a coconut for the flesh and milk
inside; perhaps a stick that enabled a bees' nest to be levered open to
get at the honey; perhaps an instrument of war; perhaps something
else entirely. What we do know is that human accomplishment in
moving from that point to what Hanson calls "substantial tool use"
was not common. In fact, it seems likely that that kind of tool use
developed once only and spread when the people who had it carried
it to people who did not – often as the result of migration, which
would not infrequently have been an act of war. Biologists therefore
say that substantial tool use may have been an evolutionary accident
of a kind that was both unlikely and – here or elsewhere in the
universe – extremely uncommon.

ONE-TIME EVOLUTION AND STEP-BY-STEP EVOLUTION

Some steps in the evolutionary process happen once and succeed quickly (that is, quickly in evolutionary terms). Hanson quotes the example of the kind of solar system capable of supporting habitable planets where life can develop. If life requires a certain kind of solar system, the success of this step happens during the formation of the solar system or it doesn't happen at all. Nothing is going to change the type of solar system later. Other steps – and Dawkins's theory about the development of the seeing eyeball is an example – can happen step-by-step, and through trial and error. This – so far as we understand evolution – is how evolution works. An organism may mutate only once in a way that produces a successful forward step. The same organism may mutate ten, a hundred, a thousand, or even more times and find that each mutation is another step down a blind alley – the changes produce no benefit and so are dropped.

It's possible to examine how life has evolved on earth, because there is a fossil record. And what that record shows is that, since Earth first came into existence, there have been five periods of about the same length between significant evolutionary changes. The first single cell fossils that have been found date back 4½ billion years which means that they came into existence less than a billion years after the earth had cooled. 2 billion years after that, fossils were being left by organisms that were still single cell but were much larger and more complex. Nearly a billion years after that, the pace of evolution can be seen to have accelerated (and the development of sex is credited with this change of speed), and now quite large fossils of multicellular life forms found. More than half a billion years after that, and you are reading this book.

There may, of course, have been lifeforms before those first single cell fossils but, if so, they would have predated the earliest known rocks in which fossils could ever be visible – and so we'll never know about them. What we do know is that the first lifeforms were coming into existence within half a billion years of the earth's cooling, and it could have been earlier. To that, we can add the likelihood that there were opportunities for life to develop at that point that would not have been available later, because the earth in its first 700 million years or so of cooling, was an immensely different place from the Earth we know today. If life had not developed on Earth at that time, could it have developed a billion years later? Possibly. And possibly not. That – the combination of environment and circumstances necessary for life to emerge – is one of the great unknowns. Paleo-biologists can theorise and hypothesise, and some of the theories and hypotheses will seem very believable (just as Dawkins's eyeball does) but proving them will probably always be beyond us.

Something else we know is that, when those large and more complex single cell lifeforms that we now see as fossils first appeared, earth had developed an atmosphere in which oxygen predominated, and what made that happen was that all the iron in the ocean had oxidised (a very slow process indeed). Those complex organisms need oxygen to breathe.

What is known as the Cambrian explosion happened about 600 million years ago and it happened at a time when the most severe ice age Earth had ever seen was ending and a supercontinent was breaking up.

And then, 65 million years ago, there was a mass extinction that saw off the dinosaurs. When the dinosaurs disappeared, the way was more open for mammals and birds to develop. And at some point, they grew large brains. How did that happen? There are many theories, but no provable facts.

And so we see why biologists – and Robin Hanson, who is an economist and not a biologist – take a sceptical view of the Drake Equation.

POSSIBLE ADDITIONAL BARRIERS TO THE COLONISATION EXPLOSION INTO SPACE

Asteroids. In the last section, we mentioned the mass extinction that wiped out the dinosaurs sixty-five million years ago. Not just the dinosaurs, but three quarters of all life on earth at the time. The cause was an asteroid. It made a crater about one hundred and ten miles wide and threw into the atmosphere such huge amounts of debris that it's believed that all light was cut off for years. We know where the asteroid hit, because the crater has been found. The evidence shows that that asteroid was about six miles wide. That's huge, and most asteroids are much smaller. But not all of them. No asteroid of anything like that size is scheduled to collide with Earth within the next twenty or thirty years, but there are asteroids out there that dwarf the one that killed the dinosaurs and sooner or later Earth is bound to take another very large hit. "Sooner or later" in astro-geological terms is measured in millions of years and it isn't likely that the great grandchildren of anyone alive on Earth today will lose their life to an asteroid collision. Nevertheless, it's going to happen at some point, it is very likely to happen before Earth has reached Stage 2 on the Kardashev Scale, and developing the technology to avoid a hit from a large asteroid, either by blowing it up before it gets here or by diverting it, would be a wise precaution.

Disease. In the fourteenth century, the Black Death killed fifty million people in Europe, which was nearly two-thirds of the population.

Today, of course, not only do we know (as they didn't then) what caused the disease and what spread it, but also we have greatly improved medical technology and antibiotics. So we could treat any similar outbreak of a threatening disease? You think so? More than eighteen million people – civilians and armed forces – died in the First World War. As the war was ending, an influenza outbreak swept the whole planet and killed anywhere between 50 million and 100 million people (we can't know precisely how many, because some of them were so far away – for example in the Amazon rainforest – that their deaths were not noticed and certainly weren't counted). Since then, we've had scares from Ebola, AIDS and other infections and they have been brought under control – so far. It would be a foolish person who gambled that no outbreak of disease would ever again overrun human defences.

War. Let's say it again: more than 18 million people died in the First World War. War is desperately destructive, and it's terribly difficult to avoid.

Those are just three of the kinds of barrier that may still exist in the Great Filter, and that humankind may fail to get past. Global warming is also a possibility. These are subjects that provoke great heat and not enough reasoned argument. Global warming (and global cooling) have always been part of this planet's ecology, and the man-made component is actually quite small. Since there are signs that Earth may be moving into a new mini-ice age, it's even possible that the man-made component in global warming may come to humanity's rescue instead of destroying it. The fact is that we don't know enough about weather cycles over the past 20,000 years to be sure.

It's worth remembering that, while the threat of any change tends to produce forecasts of disaster, a change that looks disastrous at the time can sometimes be a long-term benefit. To take one example, the Black Death. As we've said, it removed nearly two-thirds of Europe's population – but, for those who were left, it was not a disaster at all. In England, for example, there was such a shortage of agricultural labour, because most agricultural labourers had died, that those who were left were freed from the serfdom they had previously been subject to. Society needed them to become mobile, and so they were allowed to move. We've just pointed out that the part of global warming that is made by humankind may turn out to be a blessing and not the disaster it is currently being described as. The fact is that we just don't know. The data are insufficient, and positions have been taken up on both sides of the argument without enough consideration of the real evidence. "You don't understand the science" is not an acceptable response to people who question some of the stories about global warming when it comes from people most of whom – let's be frank – don't understand the science. And two noticeable facts as 2017 turns into 2018 are:

- Not only are the polar ice caps not disappearing – they're actually getting thicker; *and*

- The polar bear population is not falling – it's increasing.

As we've said, this book is not about that argument. It's worth reminding ourselves, though, that in a great many cases, we don't really know the things we think we know. What Robin Hanson and the Great Filter have to tell us is that we need a lot less position-taking and insults and a lot more careful investigation and publication of the facts.

Otherwise, the human race may have a very limited future.

CHAPTER 7

THE GREAT SILENCE

The Fermi Paradox asks to know why, if the universe is teeming with advanced civilisations, we haven't heard from any of them? It's an interesting question that has occupied a great many people for a great deal of time, but it could be turned on its head: if the universe is teeming with advanced civilisations, why haven't any of them heard from us? Because, for the most part, SETI operates in only one direction. We listen in the hope of hearing messages sent by advanced forms of life elsewhere in the universe, but we almost never send signals to those advanced forms of life. (But see Chapter 12). Suppose those other civilisations take the same approach as we do? Suppose they listen – but don't speak? Is it possible that there are observation points all over the universe where whatever happens to be the equivalent there of a human being sits at something that looks very like a radio receiver with headphones attached to whatever he, she or it uses for ears, thinking, 'Nobody is talking to us. There can't be anyone out there'?

That's one version of the Great Silence. There are others.

WE ARE NOT WORTH IT

One of the Great Silence arguments suggests that no advanced civilisation has been in touch with Earth, or even visited, because it would be unprofitable for them. We don't have anything they want. They have access to all the energy and all the raw materials they could possibly need, and as for our most advanced technology – it makes them laugh! To think that we've only just discovered 3D printing! If they'd looked at us about twenty years ago and seen the development of the first digital watches they'd have seen that we had gone from being able to tell the time simply by looking at a disk on our wrists to having to press two buttons – one to light up the face of this wonderful new device we'd developed and the second to persuade it actually to tell us what time it was. The nice, maternal ones among them might have smiled and said, "How sweet!" Those of a less understanding nature might well have felt contempt.

And so, they just don't talk to us. They have younger siblings at home, and nephews and nieces, and they know what happens if you invite someone less developed to tag along with you. You have to buy them ice creams. You have to explain to them all the things they don't understand – and the things they don't understand would fill a very large book. You have to read to them when you'd rather be playing hockey. It just isn't worth it.

On the other hand, they know we'll get there eventually. If, that is, we don't blow ourselves up first as a result of treating as a weapon to get what we want something we should have accepted as one more energy source. We'll get there, because they did. But it took them a while, and we're unlikely to be any quicker. After their civilisation had reached the level Earth is now at, it took them another fifty million years to get to where they are now, so they're not watching us with any great urgency, But they send a probe every million years or so, just to see how we're getting along.

IT'S A MORAL (OR RELIGIOUS) THING

Becoming a Kardashev Type III civilisation doesn't come easy. Yes, they're out there, but all the ones that made it almost didn't. They had to start out in the same woeful moral state as the people of Earth. If they wanted something that other people had, they "discovered" it. Just as Columbus "discovered" America and Captain Cook "discovered" Australia. There were people already there – in both places – but the discoverers deemed the discovered to be lesser beings than themselves and found for sure that they were less well armed. If they had any qualms at all, they turned to a priest who told them they were doing God's work and that God was with them all the way. Eventually, many of their descendants felt a certain guilt and shame about what had happened to the indigenous peoples. The priests, too, found that God's wishes may not have been interpreted with total accuracy. But those feelings came at a very safe time. A time when the descendants were in full control of the land, and had all the power, and could afford to feel benevolent towards the relatively few "natives" who remained.

And so it was in other civilisations in far distant galaxies. They, too, went through a very long period in which their natural or default position was conflict. All of them developed from a single pair that could be thought of as the progenitors of their species as Adam and Eve stand in for the first humans, but as they spread out across their planets, they became divided into tribes. The sense of loyalty to the tribe was absolute; the automatic approach to anyone from another tribe ranged from suspicion to outright hostility. So strong was this tendency, and so absolute in its permeation of every planetary civilisation, that it has to be assumed that tribalism is an essential and powerful driver, necessary for survival.

Something else that all of those civilisations did in their early days was to penalise half of their members for the offence of having been born into the wrong gender. Sometimes the oppressed people were "female" and sometimes they were "male," but oppression always existed. Once again, the tendency was so strong that it has to be assumed that there were good evolutionary reasons for it.

But what had once been a force for evolutionary progress came to be a brake on the same progress. It became necessary to get more people into the paid workforce than could be arranged using only half of the population, and so the other half were encouraged to leave the home and work outside it. The change was sold to them as an increase in freedom and so, in fact, it was – but what made it necessary was not (as the history books would later suggest) a desire for equality but economic necessity.

In the same way, changes in technology meant that it was no longer necessary to occupy a country rich in commodities in order to control those commodities. It was also no longer necessary to have large standing armies and navies, because some nations had developed nuclear weapons and the systems needed to deliver them to distant targets. And so the people told themselves that they had become more peaceful and more tolerant. Their grandfathers had found it natural to send armies against people who were less advanced than them, or who were every bit as advanced but had in some way upset them – they didn't share the same religion, didn't have the same political outlook, didn't agree to be dominated by some other tribe. And now, look! Those needs were gone. They only had a fairly small army, and that was focused on defence and not attack. They had become more peaceful – and therefore better people.

Of course, they did have those nuclear weapons, and the risk with nuclear weapons is that you can wipe out yourself and everyone else on the planet. And, in some cases, that's what happened, and the planet in question never got near Kardashev Scale Type III because it had destroyed all forms of life before it even quite completed Type I. But some did overcome that hurdle, and the fiction that they had developed peaceful means of cohabitation with other tribes became hardened into something approaching religious dogma. Making war was bad – no-one could argue with that. Making peace, therefore, was good – even if the price was that domination continued in a much more subtle form. Those who had power made sure that those who had none were unable to express themselves loudly enough to be heard by others who might have been affected by stories of hardship still suffered by some among them.

STAR TREK ISN'T ENTIRELY IMAGINARY, AND GENERAL ORDERS DO EXIST

Star Trek is, of course, fiction. It was made up by a bunch of writers. But history is full of ideas invented by writers that either turned out to be true or became true as technology and knowledge advanced. Star Trek looked at the question of the Great Silence and came up with the Prime Directive – The Non-Interference Directive.

That story begins about fifty years ago when the Vulcans introduced non-interference directives, but Vulcan directives were not binding on humans, and it entered the human domain thirty years from now when Starship Enterprise was exploring the planet Fazi, peopled by the Fazi and the Hipon. Seeing that the Fazi were not as developed as humans, Captain Jonathan Archer wanted to share information and technology with them. The Hipon, who were more advanced than the Fazi, asked him not to. T'Pol, who would later become a Captain in her own right but was at that time a sub-commander in Archer's fleet, advised him that the better course of action was to leave the two races on Fazi to develop at their own rate, and Archer decided that her advice was good. He felt that there should be guidelines on how to deal with less-developed civilisations (described in Star Trek as pre-warp civilisations) with whom the Starfleet came into contact.

Those guidelines were not drafted and there were a number of cases of what Star Trek authors described as cultural contamination, which did not end well. In 2165, therefore, Archer, by now an admiral, decided it was time to draw up an official code setting out what Starfleet officers should do when making the first contact with a pre-warp civilisation. Ten years later, the Council of what had become the Federation drafted a Resolution of Non-Interference, which was signed by all members of the Federation and was in full operation as the Prime Directive by the 2190s. What the Prime

Directive says is that no Starfleet personnel should knowingly interfere with the progression of a pre-warp civilisation. However, where one pre-warp civilisation is at loggerheads with another, Starfleet personnel are permitted to help negotiate agreements and compromise, but only with the agreement of both sides.

Anyone who has ever had to negotiate a family member's move to a care home is likely to understand what happened next, because care homes fall into different categories (primarily those that are purely residential and those that also offer nursing care), which means that would-be residents must be assessed. An assessment was also necessary for pre-warp civilisations. You can't assess something (or someone) until you have a working and accepted scale or set of standards; for this purpose, the Federation introduced the Richter Scale of Culture which was designed to assess the level a civilisation had reached, and to monitor its advance towards higher standards. The Richter Scale could slot a civilisation into any level from AA, which was the lowest and meant that the civilisation had developed no use of tools, to Q which was the highest (but see the next paragraph) and meant that no technology existed there which exceeded the current level of theoretical science in the Federation.

There was, in fact, an additional level – XX. To fall into that grade, a civilisation had to have developed a culture too advanced to be understood by the Federation in the Federation's current level of development.

Well, fine. But this is, as we have said, fiction. So does it have anything to say in the context of this book?

In fact, it does. What the Star Trek writers were doing was to grapple with specific moral questions that have to do with the Great Silence. It may be that Earth and its people are visible to more advanced civilisations elsewhere, and that they have moral or religious (or both) scruples about making contact. That they understand that parachuting the knowledge and technology (and mores) that come with being a Kardashev Type II or Type III civilisation into one that has not yet reached Type I could do more damage than it would do good. Or, that they understand that we are simply not ready to make use of what they offer, and that, in the present state of development on Earth, one faction or more than one faction would simply take possession of that technology and seek to use it to enslave others. And why wouldn't they fear that? It has certainly been the pattern of life on Earth in the past and no one could argue convincingly that it is not the pattern now.

So, perhaps the Prime Directive (or something very like it) is the reason why no civilisation has stepped in to help Earth progress.

The argument that will be offered against that is that, even if they don't help us, we should still be able to see them. But is that really true? Given the level of technology available to Type II and Type III civilisations, it doesn't take much to imagine that they have the ability to prevent us from seeing what they don't want us to see.

THIS IS A ZOO, AND WE ARE SOME OF THE ANIMALS

Is it possible that Earth – or, at least, the creatures on it, including humans – have been set up like samples in a laboratory by some more advanced civilisation, so that they can study what happens? A sort of cosmic social studies experiment? Yes, it is possible. It may seem unlikely, but it can't be ruled out.

If that is true, then we are being watched. Perhaps, right now, some social scientist on a far distant planet is completing a doctoral thesis with the aid of technology we are not yet sufficiently advanced even to imagine on what happens when you dump people on a planet, give them all they need to survive and progress in the way of intelligence, access to food and good atmospheric conditions, and then leave them to get on with it. It's even possible that the purpose of studies like that is to warn more advanced civilisations about what can happen when intelligent beings are left to fend for themselves and develop their own codes of conduct and systems of morality instead of having a code enforced upon them. It's also possible that a faraway alien William Golding has written a faraway alien *Lord of the Flies* describing in quasi-fictional terms the depths to which those same intelligent beings can sink in the absence of responsible moral government.

And there's another possibility:

WE WERE ABANDONED BY ACCIDENT

Adam and Eve were sent here in a spaceship more advanced than anything NASA has yet developed and told, 'Go forth and multiply. Fill the earth with your offspring. We"ll be watching, and supply ships will arrive from time to time because what we want you to do is to construct a platform we can colonise when the life-supporting days of this planet, which is far older than Earth, are about to come to an end. And remember: leave those apples alone.'

But the supply ships stopped coming. Perhaps the last arrived with a bearded man and his wife along with their children and a whole bunch of animals divided by twos – male and female. They touched down on Mount Ararat, but no one followed them. Why? There could be any number of reasons. Perhaps their home planet burned out faster than they expected. Perhaps they found another planet to colonise closer to where they were. Perhaps there was a catastrophic war that returned them to whatever their equivalent of the Stone Age was. Perhaps they were hit by a giant meteorite or asteroid and everything came to an end before they could launch their escape ships. Who knows? The knowledge would have stayed for a while, but Adam and Eve are dead, and so is Noah, so we can't ask them. They are sure to have passed on their stories, but stories become distorted and changed as time goes by. Perhaps, many years later, their distant descendants wrote the stories down, but what they wrote was coloured by their own ideas about what they wanted the result to be, and so the stories became unreliable. And then those stories came into the hands of priests and princes who had their own reasons for spreading a distorted picture of where we came from and how we should behave. One of the rules brought to Earth by those earliest settlers might have been, "Thou shalt not kill." And perhaps that idea didn't suit those in power, who added the rider, "Unless we put you in a uniform, give you arms, and order you to."

IT COULD BE WORSE

Steven Hawking said we should be glad that no alien civilisation has as yet visited Earth, and he may well be right.

Chapter 6 talked about all the unpleasant things that powerful people on Earth have done to people with less power, including enslaving and killing them. Star Trek's Prime Directive, discussed above, is the sort of code a highly moral developed civilisation might live by. But who is to say that extraterrestrial visitors to Earth would be from a highly moral developed civilisation? Is it not at least as likely that they would have acquired their current dominant position by enslaving and killing others? First on their own home planet, and then, perhaps, on others – with Earth simply being marked down as the latest target?

Enslaving and killing are two possible aims a hostile invader might have. There's a third. They might want to domesticate us. What for? For the same reason as we domesticated cows, sheep, pigs and chickens. To have on hand a ready supply of food. There's at least one song, and a great many bumper stickers, T-shirts and mugs, with the caption, *If God hadn't meant us to eat people, he wouldn't have made them out of meat*. And we *are* made out of meat, as cannibal civilisations that have existed on earth should remind us. We don't know whether the kind of meat we eat here on earth (and the kind of meat that we ourselves are made of) resembles the food eaten by extraterrestrial beings. We don't know that it isn't, either.

Now imagine that a visitor from another planet looks a lot like us. It's possible that the visitor would look at us and think, 'Well, I can't eat them, they're too like me.' It's also possible that the visitor would think no such thing. The visitor might be used to eating people that look like us on the visitor's own planet. But, leaving that aside, we pointed out in Chapter 2 that there is a very high likelihood that visitors from a civilisation on another planet would look nothing at all like us. They might look like lizards, or any of an infinite number of other possibilities. Would a lizard eat a human? Walk onto a beach where a Komodo Dragon lives and you're likely to find that the answer is "Yes".

There's no need, though, to go that far. Humans could be put to work by extraterrestrials in what would amount to a form of slavery. They could be used as we once used canaries in mines, to find out whether a place is dangerous. If the human doesn't survive, ET doesn't go there. Being exhibited as a freak in a fairground tent might be getting off lightly.

It's also worth remembering that the Prime Directive only works as long as every member of the Federation observes it. It only takes one planet – or, indeed, one visiting spaceship – to decide that the rewards from breaking the Directive exceed the risks of punishment and the Directive is dead in the water.

SHOULD WE BE HIDING?

There is no reason to assume that visitors from a more advanced civilisation will be moral, scrupulous beings concerned for our well-being. It's at least as likely that their interests will lie only in their own advancement, and if their advancement comes at the price of our freedom and, indeed, our survival, perhaps we would be wise to remember that here on Earth a horse that no longer has value for its owner goes to the knacker's yard and ends up either as glue or in a Frenchman's hamburger.

So, is it wise to try to make our presence known beyond this planet? It may not be. Our attitude to this tends to vary with the state of things here on Earth. In the 50s, for example, when the Cold War was at its height, Hollywood made a number of movies about aliens landing on Earth, and all of them assumed that the visitors would be warlike and hostile. That may have been the Western movie business transferring to alien beings the fears and distrust entertained at that time for people much closer at hand. People in the East. In the early 1980s, when fear of the communist East had to a large extent receded in the West, the movie ET portrayed a very different kind of alien visitor. But how wise would we be in assuming that the intentions of visitors from space would be aligned with how we happened to feel at the time about other earth-bound cultures? That's a question that answers itself, and the answer is: Not very wise at all.

147

Since we can't know what the intentions of alien visitors would be, perhaps the sensible approach would be not to try to make contact. But how successful would that be? We have to assume that a Type II or Type III civilisation can see us, whether we try to hide or not. But perhaps none of this matters, because:

PERHAPS THE DIFFICULTIES OF HIGH-SPEED TRAVEL MEAN WE NEEDN'T WORRY

Great swathes of thought about contact between planetary civilisations depend on assumptions about the practicality of interstellar travel. And the fact is that it may not be practical at all. We talk about travelling through wormholes and travelling at speeds faster than the speed of light – but is that something that will remain for ever a dream? We'll have more to say about this later in this book; we'll settle here for a brief introduction to the idea of FLT (faster than light) travel.

Let's begin with the kind of technology we have now. If we crank conventional rocket technology up to the highest imaginable level, it would be possible – in theory – to reach another star. Reaching even the nearest star would take hundreds of thousands of years. Reaching another galaxy would move into the millions of years. So it's impossible? Well, no, it isn't impossible. We'd either need a crew that was prepared for the fact that it would die on the way, so it would need to reproduce itself many, many times, or we'd need a crew that was prepared to travel in suspended animation knowing that no one was at the controls. They'd know at the moment of lift-off that they were never coming back, that they might never find anywhere they could settle, that their survival when they reanimated would depend on the technology that had carried the food and water with them, and that, when they got wherever they turned out to be going, not only would everyone who knew about them have been dead for several thousand centuries but that there might be no one at all still alive on Earth.

It would, of course, be essential that anyone going on such a journey should be psychologically very well-balanced indeed, as otherwise over such a period of time fights would be inevitable and might lead to the destruction of the expedition. It's also more than possible to argue that psychologically well-balanced people would never accept the conditions outlined in the previous paragraph, so the spaceship making the journey would inevitably be crewed by the differently sane.

Are there any other forms of transport that could be developed based on the physics we now know? Nuclear fuel would be a thousand times more efficient than the kind of chemical-based fuel that currently launches rockets into space from Earth. There have also been suggestions that rockets might be powered by anti-matter, but now we are moving out of the realm of the known and into the realm of the imaginable. It may happen, but not at any time soon. And one of the problems that would need to be overcome with the use of anti-matter is that the rocket itself, and all the people on it, are made of matter, and when matter comes into contact with anti-matter, it explodes.

If after accelerating for a few years, we could get a rocket close to the speed of light, then thanks to Einstein we know that time would be greatly compressed and – for the people on board – it would seem as though they had only given up, say, thirty years of their lives. That wouldn't change things here on Earth, however, and all those drawbacks connected with travelling for hundreds of thousands, if not millions, of years would still be the case here.

So what new forms of technology might be possible? And what we're really asking here is: what kinds of technology might extraterrestrial civilisations already be able to use to reach Earth?

Quantum physics makes it possible, under certain conditions, to move a quantum system from one place to another. Can this be done over long distances? We don't know. Can it be done for cargoes having significant mass? We don't know.

What about warp drive, so loved by Trekkies? Warp drive works by warping space-time. We don't yet know how to do this. We may never know how to do it. We don't know whether an advanced civilisation on a distant galaxy knows how to do it. We do know that to warp space-time would require a colossal amount of energy – the total energy stored in the Sun has been suggested. We don't have access to that amount of energy, but a Type III civilisation can access more than that amount of energy (because, if it can't, it doesn't meet the criteria for membership of Type III). What we don't know, of course, is whether a Type III civilisation exists elsewhere. We also know that warping space-time and then sending something through it would create immense tides, the force of which might very well destroy the thing (like a spaceship) that we were trying to send. And nor do we know how to dictate where the other end of the warped space-time would be.

That's a list of all the things that we on earth don't know. Something else we don't know is whether there's a civilisation anywhere that knows what we don't. It does seem more than possible, however, that – if there is an alien civilisation out there somewhere – the reason it hasn't arrived here so far and won't in the future may well be either that it can't, or that the cost in terms of damage of making that journey is too great even for a Type III civilisation to think of paying.

And that may be good news for us.

CHAPTER 8

THE EARLY BIRD THEORY

There is no mystery. The Great Filter is not something we should worry about, and the Gaian Bottleneck is the product of overactive imaginations. The fact is, it's simply too soon to be finding other advanced civilisations elsewhere in the universe. We got here before anyone else, the universe is young, we ain't seen nothing yet (and, when there is something else to see, we'll have died out anyway so it won't be us that sees it). If anyone else on another planet in a distant galaxy wants to know about Earth, they'll have to rely on the fossil record.

That, at any rate, is the view of Dr Peter Behroozi, Assistant Professor at the University of Arizona. His primary interest is in dark matter, and how it drives the creation of galaxies. He has looked at evidence collected by the Hubble Space Telescope about the birth of stars. He has also looked at planet surveys that have been carried out by the NASA Kepler/K2 mission. And what he has concluded is that, 4.6 billion years ago, when our solar system was born, only 8%

of all exoplanets likely in the course of time to become habitable wet, rocky planets existed. 92% of all the exoplanets that might one day support civilisations like Earth's had not yet been formed and – such is the longevity of the universe – most of those 92% still haven't. Time, in relation to the universe, is such a mind-boggling concept that it will be another hundred trillion years before the universe's last remaining star finally uses up all its energy and becomes cold and dead. There may well, in that unimaginable span of time, be many other civilisations that come into existence, clamber up the Kardashev Scale, and vanish. Earth's civilisation will have been one of the first, and possibly absolutely the first, to go. Why? Because we were the first to exist.

How Galaxies Form

One of the great things about the universe for those who study it is the combination of its age and the speed at which light travels. When someone on Earth looks at the moon, the moon they are seeing is the moon that existed only a short while ago. It isn't, though, the moon as it is right now; it takes light about 1.3 seconds to travel the 250,000 miles that separates the two bodies. 1.3 seconds is such a short time that it can be ignored – however the moon appears to look right now is, for practical purposes, how it actually does look right now. All the planets in our solar system are further away from Earth than the moon (which is a very good thing, as otherwise gravity would cause a collision), and so it takes light longer to reach Earth from the planets.

154

When we look at Pluto, for example, (through a telescope, of course) we are looking at the Pluto that existed more than five hours ago. An immense asteroid could have smashed into Pluto and blown it to smithereens, but it would be five hours and eighteen minutes before we saw any sign of that collision.

That's still not a very long time – but then, Pluto is only a little under 5 billion miles away from Earth, and 5 billion miles is nothing when measured on the cosmic scale. When we use the Hubble telescope to look at the farthest reaches of the universe, the light has been travelling towards us for 13 billion years. It began its journey only some 600 million years (and possibly as little as 400 million years) after the Big Bang that – so far as we are able to tell – began the universe.

The point is that it takes so long for stars and galaxies to form that those images we see of galaxies at the end of the universe are images of how they looked 13 billion years ago in the universe's very earliest stages. We can be certain that those galaxies look nothing like that today. In all likelihood, they now look very like the galaxies that are much closer to us. (Assuming, that is, that they still exist, because they may have been destroyed by some cataclysm that – because of their immense distance – won't be visible to anyone here on earth until our planet is cold and dead and there *is* no one here on earth to see it).

WHY GALAXIES MATTER

Let's not kid ourselves: the state of our knowledge about the universe is still (to put it mildly) partial and fragmentary. Humankind has been developing theories about the universe and our place in it since long before the early Egyptians and the ancient Greeks. Those theories were all state-of-the-art at the time they were developed; they arose out of what people could see and what they believed it meant, and they were slanted towards the versions that suited the religious and political powers of the time. If the Pope and the Emperor of Austria-Hungary said the earth was at the centre of the universe, it would be a brave person who proclaimed otherwise (and some were brave, and many of them died for it). Today's theories are also state of the (current) art, and they are more accurate than those of the past because we have better instruments and we've had longer to think about it, but they are by no means finalised. Because the human race is rarely troubled by self-doubt, there is a tendency to believe that those people who went before us were all morons but now, at last, we understand. Well, we don't. We are still at the very beginning of the development of a genuine understanding about this universe we inhabit, but we are starting to put some of the pieces of the jigsaw in place, and galaxies are helping us to do that.

A complete understanding of the universe would mean understanding dark matter as well as the matter we ourselves can see. It would mean understanding how things are organised on a microscopic scale and how they are organised on a very large scale. And it would mean understanding all of the laws that hold those things together. None of that would be enough, because we also need to understand not just how things are now but how they were in the past. How they have evolved from what they were to what they are. How they may evolve in the future to what they may become. And what forces are at work to promote that evolution.

We are not there yet. Galaxies can help with one of those requirements: they can help us understand how things are organised on a very large scale. And, because we are seeing the more distant galaxies not as they are today but as they were billions of years ago, we can get some idea (by comparing the galaxies near us with those at immense distances) of the evolution that has taken place.

GALAXIES ARE NOT ALL THE SAME SHAPE

Thanks to the Hubble telescope, among others, we have some wonderful pictures of what a huge range of galaxies look like. Some of them (including ours) seem to be organised in spirals, while others form ellipses. The human sense of wonder enjoys the sheer variety –

but it's also worth asking the question: why this difference in structure? Why aren't all galaxies the same?

It's possible to look at space as something completely empty. Even though the number of stars is unimaginably large, the distances between them are even larger, and it is those distances that create the emptiness. Every star, whether or not it carries with it its own solar system, is whirling through what must seem an endless empty space. And yet, collisions happen. Stars collide with stars, and galaxies collide with galaxies. The space between the stars means that it is possible for one galaxy to pass right through another and continue out the other side. Even though gravitational forces will change the structure of both galaxies, they will retain their separateness. In other cases, however, the two colliding galaxies merge with each other. Current theory says that very large spiral galaxies (once again, like ours) are formed (among other processes – this is not the only way a large spiral galaxy can come into existence) when comparatively small galaxies collide. Elliptical galaxies, on the other hand, are thought (once again, this is not the only way they can be formed) to arise from collisions between much larger galaxies.

WHAT DO WE THINK WE KNOW ABOUT THE EVOLUTION OF GALAXIES?

The important word in that heading is "think." Anyone in any doubt about the uncertainty of current theorising in this field should take a look at the last sentence in the previous paragraph. It says that elliptical galaxies are thought to form from collisions between much larger galaxies. But there is also a theory that says that elliptical galaxies were formed from clouds (clouds are next on our agenda) that did not develop a spin, while spiral galaxies were formed from clouds that did.

So: clouds. To the best of our knowledge as it presently stands, hydrogen and helium were created in the first few seconds and minutes after the Big Bang. And they formed clouds. The universe would be a very different place today if those clouds had been regular and uniform – but they weren't. What astronomers currently think is that the clouds were irregular, and some parts had greater density than others. Density brings gravity with it, and the gravity caused the clouds to begin to collapse around the denser places. Immediately after the Big Bang, the clouds of hydrogen and helium were immensely hot (less than two minutes after the bang, the temperature is estimated at 1 billion degrees Kelvin). As the clouds collapsed, they cooled, though cool is a relative term here: you still wouldn't have wanted to be out without a sun hat and some Factor 50.

This was not a single collapse – we can't see it as one huge cloud collapsing into one much smaller entity. Bits of the cloud collapsed individually to create small, dense regions which eventually became stars, while the larger but still dense regions in which they found their home became clusters of stars – galaxies, in fact. In due course (as with everything in this book, the timescales are unimaginable in human terms), the stars exploded, because that's what stars do at the end of their life. Those explosions caused renewed heat, the gas in which the stars moved absorbed the heat and the collapse of the cloud was slowed.

But the explosions did something else too: they shot carbon and nitrogen and other heavy metals into the cloud.

Nevertheless, despite all the introductions of renewed heat and heavy metals, the collapse of the gas and its transformation into stars slowed. The result was the creation – we should better say evolution – of relatively stable galaxies, though stability and galaxy are two words that should be used together only with caution.

Where do these theories come from? Simple: they come from looking at the most distant galaxies to see how a galaxy looked in its early days. And what we see when we look at those most distant galaxies is that they are small, they form clumps, and some of those clumps have very high mass and stars are being formed in those high mass areas.

SPIN OR COLLISION?

Let's look a little more closely at these two competing theories about the formation of spiral and elliptical galaxies. One theory says that, as the cloud collapsed, the degree of spin that it had or did not have decided what shape it would take. If there was a marked degree of spin, then as the cloud collapsed the spin would accelerate according to the principle of angular momentum. That increasing spin "flattened" the cloud so that it became a spiral galaxy. Clouds that had no, or only insignificant, spin at the outset did not flatten and became elliptical galaxies.

The competing theory is the one already discussed: that the shape of galaxies was decided by the kind of collisions they had with other galaxies.

While it is impossible to know which theory is correct, it is possible to outline some evidence in favour of the "collision" theory. The universe is expanding. That means that when it began it was much smaller than it is now; it also means that galaxies in the universe's early days – in fact, early aeons – were much closer together than they are now. Galaxies today are so far apart that it can be difficult to imagine them colliding with each other, though in fact they do. In those early times, it's likely that collisions happened far more frequently. Add to that the fact that there are, typically, more large elliptical galaxies in areas of space where galaxies are clustered together, which also makes collisions more likely. The final argument in favour of the "collision" theory is that spiral galaxies have much more interstellar gas than elliptical galaxies have, and it is possible to argue that colliding spiral galaxies would have set fire to a great deal of the gas that was originally there. The burning gas would have created stars, and it's possible to see signs of new stars being created in this way at this time.

It isn't possible to come down definitively on the side of one or other of these theories and it may be that there is a third theory that actually explains the difference between spiral and elliptical galaxies more accurately. If, however, the reader were to ask: Which theory does this book favour? the answer would be: We believe that the collision theory is probably closer to the truth than the spin theory. But we could be wrong.

CAN WE TELL HOW OLD A STAR IS?

Stars created more recently in astronomical time are being formed out of material that has been through a series of life cycles of individual stars. Those life cycles, as they end, eject heavy metals and so newer stars have a greater concentration of heavy metals than do those formed at an earlier stage of the universe. Spectrophotometry allows the amount of heavy metal in a star to be measured, and so gives at least a clue to the age of the star. It's interesting that some elliptical galaxies have both clusters of old stars and clusters of new stars, and comparing them allows hypotheses to be formed on how they evolved.

WHAT IS A BLACK HOLE AND WHY IS IT THERE?

There was a time, and it wasn't very long ago, when the public perception of a black hole was that it was a hostile area that anyone travelling through space would be very well advised to avoid, because a black hole was like an octopus in an undersea cave, dragging in and consuming anything that came near. In fact, there's a black hole at the centre of most galaxies including ours, and they have an important part to play in the creation and evolution of galaxies.

The black hole at the centre of the Milky Way – our galaxy – is known as Sgr A*. It is supermassive – it may be a billion times the mass of our Sun. That sounds frightening, but it is not the case that Sgr A* is capable of dragging most of the galaxy into its maw. In fact, our solar system is distant from Sgr A* by a far greater measure (in fact, some 2,000 times) than the gravitational pull the black hole exercises. But that does not mean that a supermassive black hole has no effect on its host galaxy. And the reason we are dealing with that in this chapter is that the evolution of its host galaxy is strongly influenced by the black hole at the galaxy's heart.

This book has largely ignored mathematics in general and algebra in particular. That may seem odd, given that astronomical theorists need a firm and deep understanding of mathematics, but very little can put off the lay reader as fast as a series of equations. We are going to stick with that policy, even though it means the reasoning behind some of what we say cannot be explained; those with a mathematical bent will easily find the supporting calculations online . What the mathematics tells us is that black holes needed to grow at tremendous speed, and that they grow through accretion. By accretion, we mean adding more mass, and adding it quickly – and this is where the popular idea of the black hole dragging in everything around it comes from. Accretion at that rate generates great heat, and the heat generates wind. (The heat – or at least some of it – escapes from the black hole).

The central part of a galaxy is known as "the bulge." Energy escaping from the black hole in the form of wind and heat pushes what gas remains there out of the bulge, until the mass contained by the bulge is fixed in accordance with something called the M-Sigma relation. This is part of the mathematics we promised not to burden you with; all it is necessary to say here is that there is a 1:700 relationship between the mass possessed by a supermassive black hole and the mass in the bulge around it. Take any black hole in any galaxy anywhere, and that 1:700 relationship appears to hold. The name given to the escaping wind and heat is "BH feedback," BH standing, of course, for black hole.

There are other correlations, and one of them comes as a surprise, because there is a strong correlation between the mass possessed by a supermassive black hole and the orbital speed of stars so far out towards the edge of the galaxy that – in theory – they should be immune to any gravitational influence from the black hole or, if not immune, then subject to only the weakest influence. And yet, once again, the rule holds no matter where the galaxy is: the more massive the black hole at the centre, the faster those stars on the galaxy's fringe are travelling.

We still haven't addressed the question: how did these black holes arise? And it's true that they haven't been there for the whole of the universe's life. It was many hundred million years before the first stars began to form as the clouds created by the Big Bang began to collapse. Those stars were huge, they were hot, but they didn't last very long (if we accept, as students of cosmology must, that several hundred million years is not very long). They didn't last long because, although they were packed full of nuclear fuel, they burned it at a very high rate. When almost all of the fuel had been consumed, the stars exploded. They began to shrink, which is what happens and will happen to every single star in the universe, once it has exhausted its fuel. But these early stars were hugely massive. Huge mass means huge gravity, and that unimaginable gravity meant that the core of each of the dead stars went on shrinking long after a "normal" star would have stopped (if there is such a thing as a "normal" star). They shrank, and they shrank, but what did not shrink was the power of their gravitational force – and so these early stars, as they died, formed the first black holes.

The general view of black holes is that they are destructive. And they are. But they are also creative. Visualise an unknowable beast hidden deep in a cave. People throw food into the cave to placate the beast, and the beast throws back things that the people want. All right, that's a bit fanciful – but black holes, as well as swallowing anything that gets too close, throw back high-energy particles and radiation. Those particles come in the form of jets millions of light-years long and it is those jets that – at least as seen by current theory – begin the creation of new stars. Those early black holes, then, came first, before the galaxies that now spin around them. Without black holes, galaxies would not have evolved, our Sun would not have evolved, the Earth would not have evolved – and you would not now be reading this book, because you would not have evolved, either.

Some black holes are dormant and some are active. The black hole at the centre of the Milky Way is dormant. But dormancy is not something to rely on. If anything gets too close to a black hole, the black hole will burst into life and swallow the intruder, throwing out in its place enormous amounts of energy and radiation.

WHAT HAPPENS TO THE BLACK HOLES WHEN GALAXIES MERGE?

As we've already discussed, galaxies have frequently merged in the past, are merging right now, and will go on merging into the future (though the number of mergers will fall as the universe expands and galaxies become further from each other). When galaxies merge, the black hole each has at its centre also merge, and they do so with the release of vast amounts of energy. The technology we now have has allowed astronomers to study the constellation Canes Venatici NGC 5033. It's a very interesting constellation, because there are two supermassive black holes towards the centre, one of which was probably the result of an earlier merger. What we can see when we look at this constellation is that stars are being born at a very high rate. And the most likely explanation is the turbulence created by that merger. And there's going to be another merger, because the two supermassive black holes are currently inching their way towards each other. The collision, when it comes, will cause even more turbulence than the earlier one, simply because the black holes involved are so much larger. More new stars will be created – but that won't be the only result, because the intense radiation released by the collision will cause the fabric of space-time to ripple, and scientists on earth will see those ripples and have further evidence in support of Einstein's theories.

But don't hold your breath, because that second collision is far enough into the future that it will not be visible to anyone currently alive on earth. Or to their great-great-grandchildren.

ADDITIONAL THEORIES ABOUT BLACK HOLES

As set out above, it is now clear that black holes were not some oddity as the first people to see them assumed they must be; black holes have played a significant part in the creation of the universe as we now see it. But some carry theorising on black holes further, and have begun to ask whether black holes possess the answer to a question – a fundamental question – that raised its head as soon as the big bang theory became generally accepted:

What Came Before the Big Bang?

It's a question that can't be ignored. The Big Bang cannot have arisen out of nothing. Where did all that energy come from? (A story has been making the rounds that the most sensitive instruments developed on earth recently picked up a sound from just before the Big Bang. What was the sound? Well, those who have heard it say it sounds an awful lot like "Oops." This story should perhaps be treated with a certain scepticism).

One thing we can be sure of is that the Big Bang did not come from nothing. The universe operates according to strict physical laws. We already know a great many of those laws, and the fact that we don't know all of them and don't understand some of those we do know as fully as we might (have you ever spoken to a quantum physicist?) does not change the fact that a universe ruled by laws of such strictness cannot have been created by something that flouts every one of those laws. It follows that, before the Big Bang, something existed.

A theory that has gained some support is that all of the matter in a previous universe had been absorbed into a single, universal black hole which then – having imploded to an untenable tininess – then exploded violently to create the universe we now live in. There are other theories – and this one has the disadvantage for scientists

169

that it can never be proved but also has the advantage for those who love conspiracy theories that it provides fertile ground on which to correct theories that, if unprovable, can also never be disproved. The reason this theory can never be proved lies in the nature of black holes themselves. A black hole releases energy and radiation in vast quantities. What it never releases is information about what went into the black hole in the first place. We can watch a black hole absorbing, for example, the debris from an exploding star – but we can never recover from that same black hole information about what the black hole has absorbed before we were able to watch it. When the English poet Philip Larkin was musing on the subject of death (and he rarely mused on anything else), he said:

Life is first boredom, then fear

Whether or not we use it, it goes

And leaves what something hidden from us chose

And age. And then the only end of age

That expression, "something hidden from us" describes whatever it was that created the last universe and saw it first die and then explode back into life which, this time, *was* "life as we know it." Once go into the black hole and no one and nothing will ever see you in that form again, or have the slightest idea what you looked like.

What we do know, at least according to Dr Behroozi, is that whatever caused the Big Bang did so very recently in cosmological terms. Galaxies are still being created. For all we know, before the whole process comes to an end the number of civilisations that come into existence and rise up the Kardashev scale far further than we have so far done may be very large. Why haven't we seen them? Because they haven't even begun yet. We are the first.

CHAPTER 9

NOT LIFE AS WE KNOW IT...

Ignore Dr Behroozi. There could be lots of advanced civilisations in the universe. Why haven't we seen them? Because we're looking for the wrong things. We are searching for carbon-based life forms. They may look like lizards instead of looking like us, but they are still a form of life that we can recognise. But what if they're not? What if alien life forms are machines? Robots? Beings run by artificial intelligence (AI)? And, if it comes to that, designed, developed and built by AI?

The obvious response is: How could that be? How could it conceivably be true that something as complicated as a robot could have simply evolved? Surely there must have been a human involved in its construction? And the answer to that is twofold. In the first place, it isn't necessary to believe that the robotic lifeforms in question were created and evolved from scratch. It's far more likely that they were first built by intelligent lifeforms who developed them sufficiently before themselves dying out. And in the second place, if it seems to you that a robot is far too complicated to have evolved from nowhere, how about humans? Is there now and will there ever be a robot as complicated as the human body and mind? It seems unlikely.

THE ROBOTS DESTROY THEIR MAKERS

This is not the only possibility, and probably not the most likely, but it is a fear entertained by an increasing number of earthlings and so it is worth debating. The suggestion here is that, by developing AI, humans (and any other intelligent life form that has reached this point before now or may reach it in the future) lay the foundations for their own demise. We build robots to sweep the floor, make the tea and open the door. We develop Artificial Intelligence to enable run-of-the-mill decisions to be made with no involvement of people. We put the Artificial Intelligence into the robots. Then we build on it. The robots become more and more intelligent, which means that they develop the ability to do more and more things for themselves. Robots don't procreate in the way humans do – but that doesn't mean that they can't be taught to build new robots. Ones that will (a) replace them as they wear out and (b) be better designed than the originals who "fathered" them. That's within the capacity of AI, even as it exists now, and AI on Earth is in its earliest stages. It isn't difficult to imagine a program capable of evaluating what is not working for the robot as well as it might, and designing something to do it better.

After which, the robots decide they no longer need humans; in fact, humans are a bit of a nuisance. Robots are now capable of reproducing. They've also been given sufficient intelligence to understand that they could be living lives far more rewarding than is in fact the case. It isn't hard to imagine a robot thinking, 'Is this what I was put on earth for? To make macaroni cheese for some human who doesn't have anything like my level of intelligence? And then to clean up after him? Is this really all there is?' And, from there, the robot moves on to thinking about ways of getting rid of the human in order to live the enriched independent life the robot imagines.

ARTIFICIAL INTELLIGENCE (AI)

We need to think about AI from two perspectives. The first is: where has AI, here on earth, reached? What is the present state of development? And the second is: where is AI likely to go in the future?

AI is all over the place already, and people often don't realise that AI is what they are dealing with. Amazon has Echo and the Alexa app; Apple has the HomePod. People who want to sell us things use complex algorithms that are capable of adjusting both the products being presented and the message attached to them in accordance with what they learn about our preferences and prejudices from the way we behave. And what is a self-driving car, if not an AI product? Someone has to be making decisions about where to be on the road, how fast to be driving, and how to avoid the lane-switching maniac – and if it isn't the driver, who else could it be but the car? Computers are exceeding human capability at games like Go and Chess.

It isn't just about algorithms, though. AI works through artificial neural networks (ANNs), which connect nodes in a way intended to mimic the operation of the human brain. Neural networks have become the norm in speech recognition applications, medical diagnosis, and a number of other applications that are changing human life. It isn't all plain sailing; one of the most serious weaknesses of AI has been what is known as "catastrophic forgetting." You teach a neural network to do something. Then you teach it to do something else – and the first lesson is overwritten and forgotten. That's a temporary glitch, though – IBM has already made progress in solving it.

Google has an AI group which it calls DeepMind. DeepMind showed what the future may hold when it developed a program to learn how to play the very complex Chinese game, Go. Previous game simulators weren't really AI; they used the computer's ability to store huge numbers of – for example – past games of chess in order to find the best response to a move by a human opponent or by another computer. In other words, they were taught the basics of the game – and far more than the basics – by humans. But that is not what happened when DeepMind tackled Go. The program was taught nothing whatsoever about the game. It had to learn by playing. Placing stones on the board at random and seeing what happened. At the start, it was as useless as you might expect. After only three days, however, it had understood the game and its complexities to the point where it was able to take on a grand master. They played 100 games. The program won every single one of them.

That was a very powerful demonstration of AI's ability to learn something new from the beginning. The work that DeepMind is doing on working out how proteins fold is rather more serious. It's a question that has so far defeated the best human brains. There are signs that DeepMind will provide the breakthrough.

The reason all of this matters from the point of view of this chapter is that it demonstrates that, even at this early stage, AI programs are capable of learning for themselves and developing new ideas and new ways of working. AI devices are, in effect, creating new knowledge and they are doing so by ignoring existing human knowledge. Some of the most promising results in AI, whatever the field, come when, instead of telling the program what knowledge has already been developed on a subject, the program is told: "There's the problem. See if you can find a way through it."

HOW AI PROGRAMS ARE MOTIVATED

You motivate an AI program in the same way as you motivate a dog you are training. You offer it rewards – treats. In the case of the dog, the treat might be a special kind of biscuit, or a chocolate drop formulated especially for dogs. (Please don't give your dog the same kind of chocolate you eat yourself. The dog will like it, but it won't be good for the dog). Right now, you motivate an AI program by telling it what it should regard as a treat. In effect, you say, 'If you reach this point, you are succeeding in solving the problem. And if you reach that point, you have succeeded in solving the problem.' You have made solving the problem the program's reward. That suits you, because the program has been motivated to deliver the results you are looking for. Will that still be the case when AI programs are providing their own motivation? Possibly not. For the purposes of this chapter, what happens when an AI program realises that – from its perspective – the best outcome is not the one proposed by the human who programmed it, but one it has devised for itself? One that might even be damaging to humans?

Science fiction writers were the first to address that question.

THE LAWS OF ROBOTICS

Isaac Asimov foresaw the concerns we just outlined and developed a series of laws for robots that would prevent them from acting in a way hostile to humans. There were originally three of these, but he increased it to four:

- A robot may not harm humanity, or through inaction allow humanity to come to harm

- A robot may not injure a human being or, through inaction, allow a human being to come to harm.

- A robot must obey the orders given it by human beings except where such orders would conflict with the First or Second Laws.

- A robot must protect its own existence as long as such protection does not conflict with the First or Second Laws.

That sounds like protection, but there are some problems with these laws. The first is: we already set laws in profusion to govern the conduct of humans, and they are regularly broken. What confidence should we have that laws for robots would be any better observed? Of course, against that can be set the argument that it will be us – the humans – who programme the robots and we could ensure that they were programmed to obey the laws we set for them.

But could we really? What happens when the artificial intelligence with which the robots are programmed works out that the robots' interests would be better served by setting aside the instructions of humans?

In any case, how likely is it that all AI robots – or all non-AI robots for that matter – will have these rules built into them? Some of the biggest funders of AI and robotic development are military: the US army; the Russian army; the Chinese army; and no doubt most other armies of any size. A lot of those robots will be required to do jobs currently done by soldiers. Do armies tell their soldiers, "You may not injure a human being"? Of course they don't – there'd be no point in having soldiers if they did that. So why would we expect that they would instruct their robot soldiers that way?

Even robots programmed for non-combatant roles are unlikely to have the laws of robotics built in. Take, for example, the bomb disposal robot. Having a bomb decommissioned by a non-human is a potential human lifesaver. But what happens if, while the robot is dismantling the bomb, a soldier from the side that planted it interferes? Will the robot simply accept that the job can't be completed? It's hardly likely, is it? At the very least, we can expect the laws to be modified so that, instead of simply saying, "You may not injure a human being," they will say, "You may not injure a human being of your own side."

OTHER KINDS OF ROBOT

The word "robot" calls to mind a mechanical device. But that does not cover all robots, and it certainly doesn't cover all robots currently being conceived and designed. It's possible, for example, to create a robot from proteins and DNA and to use it for the correction of gene disorders. Those robots will be injected into humans and will be expected to integrate with the body into which they have been injected. Will they be programmed with Asimov's Laws? How will that work? Which has first call on the robot – Asimov's Laws, or the information and instructions from the genome?

OTHER KINDS OF HARM

Robots are becoming much less like androids and much more like people. They are also learning about emotions. What happens when a robot falls in love with a human being? Will it use its artificial intelligence in the same way as humans sometimes do to work out ways of winning the human's heart at the expense of a rival? Perhaps even breaking up a marriage in the process? Doesn't that count as harming a human? Even if they don't actually go whole hog and murder their love rival?

And what about the harm to humans who fall in love with a robot and find that the robot rejects them? Will robots have to be instructed to yield to the love of a human? Okay: so what happens when two humans fall in love with the same robot? Or three? Or more? Someone is going to be two-timed – isn't that harm?

THE ROBOTS' MAKERS DESTROY THEMSELVES

We began this chapter by positing that robots might destroy their makers (us). The idea, and it is a common one, was that robots, imbued with artificial intelligence and having formed a view about what is in their interests that clashes with the interests of the humans who built them, will turn on their makers and destroy them. Those humans knew that this was a possibility, and so they programmed the robots with an updated version of Asimov's robotic laws to prevent it – but it didn't work. Perhaps the laws weren't up to the job; perhaps they weren't very well programmed; perhaps they simply weren't strong enough to overcome a level of robotic self-interest that had been learned from their human makers. If Man makes robots in Man's own image, we should perhaps ask whether that is really a good idea, given the history of mankind's behaviour towards each other.

The fact is, though, that this is not the most likely outcome of the development of robotics and artificial intelligence. What is more probable is that humans will continue in their present self-destructive trajectory, and intelligent life on Earth will be brought to an end as a result of nuclear war, the escape into the environment of lethal man-made pathogens, or some other disaster created by humans. Or perhaps we'll just be hit by a huge meteorite which will cut off all access to sunlight for enough time that we all starve to death.

And that is the end of human civilisation on Earth – indeed, the end of the human race. But the machines will still be there. They may have to develop ways of dealing with potentially lethal radiation, if that has not already been programmed into them, but they will exist. And the artificial intelligence that programs them has reached a sufficiently advanced level that they can (a) realise what has happened and (b) decide to carry on on their own.

In this case, the machines have some advantages. As TS Eliot wrote, "What is only living can only die." But the machines are not alive, or at least not in that sense, and they don't die. They may corrode; metal fatigue may cause them to become impotent; their circuits may burn out – but they are not subject to a finite lifetime in the way that humans are. And, because of the level of AI they enjoy, they can replace themselves. They have the potential to go on forever, whatever "forever" may mean in a universe that began with a singularity and will probably end the same way.

There's more. As long as they have the ability we have already discussed to protect themselves against radiation, space holds no terrors for machines. They don't need breathing apparatus. They don't eat, so they don't need to take food supplies with them when they travel and nor, on very long journeys, would they need to develop vessels with what amounted to farms on board. If they assemble the right robotic crew, including robots with all the skills they are going to need, they can land on passing asteroids – or, indeed, other planets – in order to mine the materials to replicate and to repair both themselves and their spacecraft.

Nor is time a problem. Persuading someone on earth to board a spacecraft that intends to travel for thirteen billion light years to the far end of the universe just "because it's there" means that the crew has mental health issues from the very start. Not so with robotic space travellers – they (or their self-produced replicas) are going to live forever, and forever includes thirteen billion light years.

So, this is the more likely reason why robots from Earth may be travelling through space long after all intelligent life (and possibly *all* life) has ceased to exist on the home planet. Not that the robots will rise up and put an end to humanity, but that humanity will manage not to put an end to itself until it has created a level of artificial intelligence in robots that allows them to pick up the baton.

HACKING

Hackers are already using artificial intelligence in order to steal vital data. Everyone has read stories about hackers stealing databases belonging to major brands. The databases often contain personal details about the brand's customers – names, addresses, passwords, purchasing record and, if the customers are really unlucky, their bank and credit card details. 'But don't worry,' say the owners of the hacked databases – 'all of the information is encrypted, and to find the key by a blunt force attack (that is, by trying every conceivable combination of characters and symbols until the cryptographic key is found accidentally) would take trillions of years.' And that may well be true. But it doesn't provide much comfort, because hackers using AI have developed ways of downloading that key in one second (for native systems) and three seconds if the database is in the cloud.

How do they do that? By what are called side channel attacks. The hackers know that the cryptographic key is being called in the system with great frequency. It has to be, or no new data could be encoded. So, hackers use a program called "Flush+Reload" to empty the processor's three levels of cache, knowing that they have to be instantly refilled, and they record the data that is stored there, and then they use AI to identify those calls on the processor likely to be loading the bits of the key. Within those one to three seconds, if the system is not running adequate protection, they have the key (and that adequate protection is also a form of AI, because it relies on spotting program elements that should not be running).

That's a nasty thought for anyone who has personal information stored on any database, anywhere. And that means almost the entire population of any advanced country and many that are less advanced. But there's an even nastier thought. All modern armies run on IT. They transmit orders and information online. They use cryptographic keys to do so. The side channel attacks we've just described are not used only (or indeed mainly) to get hold of the personal details of people who shop online. They are used more regularly and more frequently to steal confidential information from manufacturing companies, including those involved in making armaments.

So here we have a scenario in which one country can interfere with the instructions being sent by another country in relation to its use of weapons. Imagine if every nuclear bomb in every silo in one advanced country received an instruction to explode. Right where it was. There and then. And then imagine the country that had been hacked in this way retaliating by sending a similar instruction to every nuclear bomb in every silo in the country that had hacked them. Mutually Assured Destruction (MAD), the policy that has – in all likelihood – prevented the nuclear powers of this world from blowing themselves and the rest of us to kingdom come, is also the policy that most people rely on to say, 'Don't worry about nuclear war – it's so frightening that it will never happen.' The sequence of events we just described would have drawn the teeth of MAD, creating a situation in which intelligent machines might very well be left unattended to make their own way in the universe.

THE INDUSTRIAL INTERNET OF THINGS

The Internet of Things has received a lot of publicity. But what is of more interest is the Industrial Internet of Things (IIoT), because the great benefits won't come from allowing people to open their garage door as they approach it, or to make sure that they dinner is cooked at the time they want to eat it, even though they are currently somewhere else. The great benefits will come (and are coming) from such applications as putting process industry online at the point where manufacture is actually taking place.

It's been possible for many years to embed sensors in industrial processes. The software has existed for an equal number of years to analyse in real time what was actually happening inside the oil refinery, the food processing plant, or the steel mill. What was not possible until fairly recently was to transmit the data in real time from refinery, plant or mill to a central point where it could be analysed. Analysis had to be conducted on-site, on a piecemeal basis, which was unsatisfactory for a company that had a number of sites and wanted to look at the whole picture at the same time. The Internet filled that gap. Now the Industrial Internet of Things allows central monitoring of everything that's happening. It provides an enormous improvement in efficiency as well as reducing costs. Unfortunately, it also opens the processes to observation by hackers.

Some observation is just that – observation. A geek in a bedroom getting kicks out of seeing what is happening in a steel mill on another continent. But some isn't. Some hacking is looking for blackmail opportunities. And some is prepared to exploit vulnerabilities in security to interfere with the process in some way. Carried to its extreme, interference can mean: poisoning the food that is being produced; rendering the steel unusable; and dumping the oil in a way that pollutes groundwater.

The lesson is: as technology becomes more complex, it also becomes more vulnerable to hacking. And some hackers have the worst possible motivations. Hacking could well be what leads to the extinction of human life on this planet.

ELECTRONIC LIFE FORMS

So far, this chapter has explained why Lord Martin Rees has said that, "The existence of life away from Earth does not necessarily mean that there is intelligent life there. My guess is that, if we do find intelligent life, it will be an electronic entity of some kind." Lord Rees is British Astronomer Royal and a professor of cosmology and astrophysics at Cambridge. He thinks – and, once again, the reasons for his beliefs are set out above – that AI could enable machines to dominate the universe. The future, he says, is not in intelligent life on its own but in "electronic intelligent life."

Rees does not believe that there will ever be meaningful travel of human beings in space. Not even, to any extent that makes it worthwhile, within our solar system. He believes that humans will always be earth-bound. Tied to the earth. That, if any long distance, long-term manned expeditions leave Earth, the "men" in question will be machines.

This does not necessarily matter. What are the reasons for interstellar exploration? If it is merely to gain information, then machines can do that for us and send information back to earth. The timescales will be such that no one who sent the expedition on its way in the first place will still be around to see the results, but so long as humans remain on Earth, and provided that they have not sunk into a new age of barbarism, they will still be able to benefit. And then there's energy. Everything that has ever been achieved, whether on this planet or, conceivably, any other, has needed energy. The most potent source of energy in the Milky Way is the black hole at its centre. Machines can be sent to exploit that energy by collecting it and channelling it back to Earth. It's possible to imagine a number of other scenarios in which people on Earth benefit from the space travel of machines. But it's also possible to argue that none of that really delivers what humans want, which is that it should be *they* who travel, *they* who explore other civilisations, and *they* whose name becomes associated with the voyages in the same way as the names of Columbus, Marco Polo and Captain Cook. It isn't necessarily logical; it *is* human.

HUMANS IN THE MACHINE

And so it is that a good deal of theorising as well as moviemaking and the writing of science fiction imagines the transitioning of real people into the machines. This is, quite frankly, imagination bordering on the fantastic. But that's exactly what they said to Galileo, so let's not write the idea off entirely. In this version of the future, humans have realised that their only hope of sustaining their race into the future is to become part of the machine. They'll have no blood, flesh and bone components. No legs, arms, stomach or head. None of the more usual attachments for enjoyment. What they will have is the essence of their own brain transformed into a neural

190

network inside a robotic life form. Jacob Smith will still be Jacob Smith – if we accept (and why wouldn't we?) that the essence of Jacob Smith is not in the particular shape of his upper lip or the length of his index finger, but what goes on inside his head.

And so, intelligent life forms would be able to travel great cosmic distances by anchoring themselves in mechanical bodies that do not wear out. Who knows? After three or four billion light years have passed, Jacob Smith might meet the representatives of an advanced civilisation in a far distant galaxy. Will he think it worthwhile? That must, of course, depend. It may be that Jacob will experience a meeting of minds that gives him the answer to the greatest existential questions of all which, contrary to popular opinion, do not include "When the doorbell rings, why does the dog always think it's for him?" The questions we have in mind here are: Why are we here? And who sent us? On the other hand, he may look at the graffiti on the subway wall and think he might as well have stayed at home.

Or, if the people in that civilisation are anything like the people here on Earth, they may simply shoot him.

CHAPTER 10

A LONG ROAD AHEAD OF US

Calm down, already. According to Leo Tolstoy, the two most powerful weapons life has to offer are patience and time – and we haven't given the search for extraterrestrial civilisations anything like enough of either.

There is a way of looking at humanity's place in the world that illustrates graphically something we all know but mostly fail to understand. It uses a 24-hour clock to show the history of Earth since it came into being. In other words, Earth's entire history is illustrated as a single day. And where during that day do we – Homo sapiens – turn up? Humans arrived on the scene a little more than a minute before the clock strikes midnight.

But in fact the timescale is much shorter than that. The creatures who became us first walked on earth about 6 million years ago, but the earliest recognisably human form was seen only about

200,000 years ago. The earliest civilisation, at least as far as we know, was in Mesopotamia. That's a word from the ancient Greek, it means "the land between rivers," and it covered roughly the area now occupied by Iran, Syria and Turkey. The earliest signs of that civilisation were in 3500 BC, which is less than 6000 years ago.

And so, in fact, no one capable of observing signs of a civilisation on a distant planet existed until a lot later than one minute before midnight on Earth's 24-hour clock. And it happened even later than that suggests, because the people of ancient Mesopotamia made all kinds of observations about our place in the universe but none of them entertained the idea of habitable exoplanets. That would not become possible until civilisation had become entirely industrialised; although people were making things out of bronze more than 4000 years ago, that was a very localised, small-scale version of industrialisation. The first iron bridge in the world was built in 1779 in Ironbridge in Shropshire, England. Ironbridge is in a valley of the River Severn called Coalbrookdale, and there's a hint right there of why Ironbridge calls itself the cradle of the Industrial Revolution. There was water power from the river, there was coal and there was iron ore, all locally available.

So, it was possible before 1800 for educated people to study extraterrestrial civilisations? Well – no. It was possible for educated people to imagine that extraterrestrial civilisations might exist, and many did, and in some countries this caused a clash with both civil and religious authorities that led to their death if they refused to recant from such a ridiculous and ungodly notion. But see them? Hear them? Know for certain that they were there? How would they do that? Refracting telescopes first appeared early in the seventeenth century in The Netherlands. Sixty years later, Isaac Newton produced the first reflecting telescope. By the end of the nineteenth century, telescopes had improved enough for Italian astronomers to be able to observe canals on the surface of Mars. And that, of course, was something of a problem, because there are no canals on the surface of Mars. We've been there now, and we know that for certain.

It's only in very recent years, with the development of such spectacular instruments as the Hubble Space Telescope and the rocketry necessary to hurl them into space, that humankind has been able to look closely at the surface of some of the nearer planets. As well as explorations of the moon and Mars, NASA has conducted flybys of Jupiter, Saturn, Uranus, Neptune, and Pluto. Those visits have provided astronomers with a great deal of information about the outer planets of our solar system – among which is the fact that there is no visible sign there of any advanced civilisation. (There's no sign of Fred Flintstone, either). Will an interstellar mission one day send back pictures that provide evidence of civilisation in some distant part of our galaxy or some even more distant galaxy? Well, perhaps. But the realities of cosmic travel mean that, by the time the pictures get here, it's very likely that there will be no one left on earth to study them.

All of which could be very depressing, were it not for the existence of radio telescopes.

If we were able to speak to those early civilised humans in Mesopotamia, they would tell us that anything that could be observed in space (not that they saw it as space) would be observed with the eyes. It would be visible light that made it possible for the object to be seen. There wasn't anything else. Was there? Of course, by the time Isaac Newton came along, humans were much more educated. Possibly so – but Isaac Newton would also have believed that visible light was the only way to see a heavenly body; you certainly wouldn't be able to hear it. Newton knew, as all educated people knew by that time, that light and sound both travel in waves, but they are different kinds of waves. Visible light is a part (a very small part, as it happens) of the electromagnetic spectrum and is able to travel through a vacuum. Sound is a different kind of wave, not part of the electromagnetic spectrum, and it needs a medium to travel through. The medium doesn't have to be air, though air is the one we are most familiar with, but it has to exist. And space is empty, and by the time of Isaac Newton, we knew that. Newton's successors, the great scientists of the eighteenth, nineteenth and early part of the twentieth century knew the same thing. In space, no one hears you scream – because sound can't travel through a vacuum.

And yet, we now know that we can "hear" astronomical objects as well as see them.

Like so many things – like, for example, Post-it Notes and Viagra – this is something that was first noticed by somebody who was looking for something else. In 1932, at the Bell Telephone laboratories in New Jersey, an engineer called Karl Guthe Jansky was working on a project to make the sound emitted by radio and radio telephony receivers clearer. They were constantly interrupted by static, and Bell had told Jansky to find out where the static was coming from, so that it could be dealt with. He put together an array of reflectors and dipoles designed to receive shortwave radio signals and mounted it on a turntable that could be rotated so that the antenna could be pointed in any direction in order to pinpoint the source of the interference. This was no mini-construction; it was a hundred feet across and twenty feet high, and in a nearby shed was a recording system that drew graphs based on what was received.

Jansky ran his apparatus for months, painstakingly recording all signals received from every possible direction. At the end of that time he had enough data to say that there were three kinds of static. There were nearby thunderstorms. There were distant thunderstorms. And then there was another kind – a hiss, faint but steady, that repeated on a twenty-three hour and fifty-six minute cycle. That cycle is exactly the length of the time it takes any "fixed" object to return to its original place in the sky – it's called the sidereal day. That was the clue Jansky needed to hypothesise that the hiss was not an earthly sound but was coming from somewhere outside our solar system. Checking his recorded observations against astronomical maps led him to conclude that the origin of the sound was in our galaxy and was, in fact, near the centre of the galaxy in the constellation Sagittarius.

Although Jansky probably could not know this at the time, he had made an enormous leap forward in what would become the search for extraterrestrial life. Five years after Jansky had begun his search for the sources of static, an amateur radio enthusiast in Illinois called Grote Reber built what is thought to have been the first radio telescope. And that date – 1937 – can now be recognised as the first time it would really have been possible to detect signs of extraterrestrial life in the way that, for example, SETI is now attempting to do. Eighty years ago. A small fraction of a second before midnight on Earth's 24-hour clock. Is it any wonder that some commentators tell us we have not looked for long enough, and that we are expecting results far too quickly? It isn't a question of "Where are they?" but "When will we first hear them?"

Grote Reber's early radio telescope was a parabolic dish, thirty feet across. It was in his backyard, and although there is no record of how big the backyard was, that telescope must have taken up a lot of it. He confirmed Jansky's conclusion that the first extraterrestrial source of radio transmissions was in our galaxy. He then surveyed the sky looking for other sources of high-frequency radio transmission, and found a number of them. So far, so hobbyist – but then came World War II and the development of radar that the war made urgent. This delivered the technology that, after the war, enabled radio astronomy to go forward faster.

It was now apparent that the electromagnetic spectrum is much longer than simply the small portion that delivers visible light that can be seen by the human eye. As seen by animals and birds, the visible light part of the electromagnetic spectrum is larger than humans had imagined, because birds and animals can see things with their eyes that we can't see with ours. The spectrum is, however, much more extensive than that; gamma rays are at one end and are transmitted over very short wavelengths, radio waves are at the other end with long wavelengths, and in between those two extremes in increasing order of wavelength are x-rays, ultraviolet light, visible light, infrared radiation, and microwaves.

Many objects in the sky emit visible light, so that we can see them, with the aid of a telescope if they are sufficiently far away, but they also emit other radiation. Planets, stars, galaxies, nebulae, pulsars and quasars all emit radiation over a large part (and sometimes all) of the electromagnetic spectrum. NASA and other bodies have technology that allows them to represent data received as invisible radiation into a visible form, and many of the "images" from the Hubble Space Telescope that have been published by NASA were, in fact, received as non-visible radiation and converted into visible form so that we can see and admire some quite amazing pictures.

Radio telescopes have, in fact, supplanted optical telescopes as the main source of astronomical information. They have the advantage over optical telescopes of being available for use twenty-four hours a day, and not just during the hours of darkness. Because a great deal of the radiation received from distant radio sources is weak, the telescopes need to be extremely sensitive and are often joined with other telescopes, sometimes a considerable distance away, in an array that makes it possible to magnify the "images" received. They are also usually sited some distance from populated areas in order to avoid electromagnetic interference created by man-made electronic devices like television, radio, and cars. Some radio telescopes have extremely large parabolic dishes, but the majority resemble the antennae used for receiving television signals, though they are much larger. The dish shape is mostly used for short wavelength signals and the ratio of the dish's diameter to its angular resolution is what dictates the particular wavelength that a dish can receive.

The Atacama Large Array shows exactly what can be achieved when building a radio telescope. It was completed in 2014 and is already transforming what is known about the universe. It's in Chile, but it is more than simply a Chilean venture, because the nearly $1.5 billion it cost to build was contributed by the USA, Canada, Chile itself, and countries in Europe and Asia. It is the most complex telescope in existence. It illustrates why the best telescopes tend to be arrays rather than single huge devices: gravity would make it impossible for one large device to achieve Atacama's amazing stability.

Atacama – or, in fact, ALMA – has sixty-six antennae spread over a distance of almost 10 miles. When observing at the smallest wavelengths, it has a resolution ten times that of the Hubble Space Telescope. The surface of each antenna is tuned by hand for the most accurate reflection of light waves only 400 micrometres in length. If you wonder what that is, it's about the amount a human hair grows each day. Why do we mention human hair? Because, in any dish, a bump larger than one third of a human hair's diameter would scatter the cosmic waves the antenna is there to capture.

There's a problem. The wavelength of some of the signals being received by ALMA is so short that roughly 1 trillion arrive each second, and we don't yet have the computer technology to handle that sort of data stream. But radio astronomy is at the very leading edge of technological development, and the guys behind it came up with a solution: signals are received by ALMA and as they leave it, they are mixed with a carrier wave of longer wavelength.

There's another problem. ALMA is completely dependent on its electronics, and electronics develop their own signals. The answer to that is to keep the receivers in each antenna as cold as possible and that is done using liquid helium very close in temperature to absolute zero.

RADIO INTERFEROMETRY

Radio interferometry was developed immediately after the Second World War. It is the technique mentioned above of connecting a number of antennae to act as one much larger antenna in what is known as an array. An array may be a collection of parabolic dishes or of one-dimensional antennae, or it may be a two-dimensional array of dipoles which are themselves omnidirectional. The telescopes in an array may be connected by coaxial cable in the way that computer terminals that had no intelligence in themselves were at one time connected to control units, with the control units themselves talking to mainframe computers or mid-range computers over other forms of transmission. They may also be connected by optical fibre or by a number of other types of transmission line. More recently, the technology has been developed to allow signals recorded at unconnected antennae to be transmitted to a central processing unit where they are combined; this combination can be and usually is carried out at a time subsequent to the signal collection and is known as VLBI (Very Long Baseline Interferometry). Although there is no doubt that interferometry increases the total of collected signals, that is not its main purpose. The main benefit that radio interferometry gives is the ability to greatly improve resolution of the signal by what is known as aperture synthesis. Signal waves from the various telescopes are superposed on each other in a process known as interference, which is where the name interferometry comes from. The result is to simulate a telescope with a resolution equivalent to what would be gained from a single antenna with a diameter of the distance between the two furthest separated antennae in the array. Some Very Large Arrays are therefore able to "see" the source of the transmission with an almost unimaginably high resolution. Work continues to improve resolution; an array currently under construction in Europe comprises some 20,000 antennae, none of them particularly large but spread over forty-eight different locations with a diameter of several hundred miles.

It is in this way, for example, that detailed images have been obtained of aspects of the Cosmic Microwave Background left over from the Big Bang and now providing undreamed of information on what happened in the Universe's earliest moments. ("Earliest moments" in this context means the first few million years).

WHAT ARE RADIO ASTRONOMERS LOOKING FOR?

If everything we just outlined makes it sound as though the search for extraterrestrial civilisations has become easier, we've misled you. It is still extremely difficult, not only to know what to look for, but to know where to look.

In some ways, the very power of radio astronomy increases rather than reduces the difficulty. Radio astronomy makes it possible to see gas and cosmic dust that would otherwise be completely invisible. That ability is of vital importance in unravelling the mystery of how the Universe evolved, what happened after the Big Bang, and where we may be going. It helps prove or disprove the theories of Einstein, Hawking, and other leaders in cosmological theory. So what is more important – to look for a branch of Starbucks on a planet in some galaxy light years from the Milky Way, or to test some of the most advanced theory? Many would say that both are of equal importance, but budgets are not unlimited and nor is the amount of time scientists have available to study collected data.

If anyone out there is sending signals to look for a civilisation like ours, what wavelength are they transmitting on? We don't even have an answer to that question, and so we have to cover the entire spectrum. And that may take a very long time.

WHAT HAVE RADIO ASTRONOMERS DISCOVERED SO FAR?

We'll start with something close to the purpose of this book. In 2008, radio astronomers found molecules of hydrogen cyanide and methanimine in a starburst in the constellation Serpens. Those are "prebiotic" molecules – they are among the building blocks for life. That means that we have found a collection of stars in our own Milky Way galaxy that appear to have reached a stage in the process on the way to the generation of life. Will they get there? Will an exoplanet (if there is one) in Serpens one day have sentient beings peering at the data from a radio telescope of their own that has scrutinised what remains of Earth, scratching whatever they have where a chin should be and murmuring, 'Hmm. I wonder'? There's no way of knowing, and prebiotic molecules are such an early stage in the formation of life that the answer will not be forthcoming until long after the Earth is dead. The point, though, is that radio astronomy has made it possible to find a place a long way from here in which at least some of the conditions for life to form exist.

And then there's ALMA. We've already described ALMA; its purpose is to collect the weakest, shortest wavelength signals. And that makes it possible to see things and to understand things that transform thinking about cosmology. ALMA has found glycolaldehyde in the star system IRAS 16293–422. Glycolaldehyde is a simple sugar and one of the building blocks for RNA; it gets genetic information into cells. Bear in mind that ALMA has only been in operation for about four years – we have to ask just how many star systems are going to turn out to possess the raw materials for life. IRAS 16293–422 is a young star system, a long way from the maturity of our solar system. It's far too early to say whether an exoplanet in that system will one day host intelligent life, but it's increasingly clear that the raw materials for life are far from unique to Earth.

Other radio astronomy discoveries include greater accuracy in understanding what is happening on nearby bodies. Mercury, for example. Mercury's orbit round the sun was well known to be eighty-eight of our Earth days. But in 1964, the radio astronomer Gordon Pettengill studied the data from the Arecibo radio telescope and realised that this calculation was wrong. In fact, the orbit is fifty-nine

Earth days long. That may sound like a small thing, but in fact it h
great significance, because exploration of space, whether by going
there, sending an unmanned craft, or simply observing, requires
absolute accuracy.

Then there's the question of what asteroids look like, and here
we are once again in Arecibo. In 1989, the Arecibo telescope
collected the signal data than allowed radio astronomers to create an
image – a three-dimensional model – of the asteroid 4769 Castalia. (If
you're wondering, it looks like a peanut. But a peanut large enough to
destroy all life on earth should it hit us – and that makes knowing as
much as we can about this and every other asteroid vitally important).

A pulsar is a "white dwarf" – a highly magnetised, rotating
neutron star that emits electromagnetic radiation in the form of a
beam. Without that radiation, radio astronomy would never have
discovered pulsars. This was considered important enough that two
astronomers were awarded the Nobel Prize for Physics for their
discovery in 1993 of the first binary pulsar (one that has another
white dwarf or neutron star close by mediating the pulsar's
gravitational direction and mass). Ten years before that, five other
radio astronomers had discovered the first "Millisecond Pulsar" –
one with an extremely rapid rotational period. In fact, that particular
pulsar spins 641 times every second. Without radio astronomy, we

.s   : known that they existed; now, more than 200 have

ιolecular cloud is a collection of gas and dust that, like

um, is very close in temperature to absolute zero. The

.at form it have been invisible, partly because they lack the

to send a signal of any strength, and partly because they are

.red by dense cloud. But ALMA can see them.

Some of the answers we've already given in this book to questions like,

- How do stars begin to form? *and*

- How do stars stop accumulating when they're big enough?

came from ALMA. Something very interesting that ALMA made clear is that not all of the dust and gas being sucked in by a planet during its formation remains with the planet; some of it goes straight past and falls into the star. What that means is that star formation and planet formation work in tandem – one feeds the other.

Other interesting observations by radio astronomers:

- Rho Ophiuchi 102 is a brown dwarf surrounded by a disk of particles that could well be forming a planet very like Earth

- Supernova 1987A was once a supergiant star until it exploded in the Large Magellanic Cloud. Watching it has provided many astronomers with a course in how stars evolve. What they didn't know until ALMA came on the scene was just how much carbon is forming. If this continues, to the point where the amount of carbon produced is more than the textbooks say should have been possible, we won't be able to dismiss the evidence – it's the textbooks that will have to be rewritten

And so one look at a fairly restricted area of cosmology, and one that has been in operation for a very short time, shows us just how much we still have to learn. We haven't heard from any other civilisation in the universe, but that doesn't mean they aren't out there. It may simply mean that we haven't been looking long enough and hard enough. As Tolstoy told us, we need more patience. And more time.

CHAPTER 11

IN A GALAXY FAR, FAR AWAY

Perhaps there is an advanced extraterrestrial civilisation somewhere out there. Perhaps there are many. And perhaps they are so far away from us that we lack the technology to see them, whatever "see" may mean in this context, and any signal they might have sent in our direction has such a distance to cover it hasn't reached us yet.

The universe is big enough to make this entirely possible. And not just this universe: Steven Hawking, who went to another dimension just before this book was finished, leaving an unfillable black hole behind him, hypothesised that the Big Bang might have been the beginning of not one but many universes all of which shot off in different directions. Perhaps one day we may have the technology that enables us to receive and decode signals from another universe entirely. We certainly don't have it right now. And nor, really, are we capable of receiving signals even from a civilisation in a distant part of the universe we live in. Even receiving a message from a civilisation at the far end of our own galaxy might prove testing.

So how big is the Milky Way galaxy where Earth has its home, and how big is the universe as a whole? And, as it's been expanding ever since the Big Bang, doesn't it – like the little girls Maurice Chevalier sang about in those innocent days when it was still possible to mention such matters – get bigger every day?

HOW BIG IS OUR GALAXY?

Even calculating the size of the Milky Way has proved difficult over the years, but we are now pretty sure that it has a diameter somewhere between 100,000 and 150,000 light-years. We are some distance along one arm of the Milky Way, but by no means at the end, so let's say that a signal travelling at the speed of light from the furthest reaches of this Galaxy would get here about 1000 centuries after it had been sent. As explained in Chapter 10, we have really been effectively looking for such signals for a very short time and we are as yet by no means covering the whole range of possible wavelengths and sources when we do look, so it's entirely possible that such a signal might not as yet have been noticed.

When we want to calculate the size of the whole universe, it becomes significantly harder. It's also necessary to distinguish between the universe and the observable universe, because there is more universe than we can see.

The process astronomers go through to calculate just how big the universe is is known as the cosmic distance ladder. It has that name because there are a series of steps – rungs on the ladder, if you like – that have to be gone through, depending on just how remote is the part of the universe whose distance from here we are trying to measure.

Rung one is the measurement of the distance of the various planets in our own solar system. And that's easy, as these things go, because all we have to do is bounce a radio wave off the planet in question and see how long it takes to return to us. We know how fast the radio wave is travelling; therefore, we know how far away the planet was when the wave bounced off it.

That gives us a very precise measurement. But it doesn't work so well for objects outside our solar system. So now we have to go up one step on the cosmic distance ladder, to parallax.

Try this. Look at an object in front of you — but look with only one eye. Now close that eye and look at the same object with the other eye. The object has moved its position. Not much, because your eyes aren't very far apart, but it has moved. That movement is called parallax. And you can use the distance between the two instances of the object as seen from your two eyes to calculate the exact distance the object is away from you.

You could probably have done that more easily with a tape measure. But a tape measure won't tell you the distance to a star in another galaxy. For that, you really do need parallax. And you need a bigger baseline than the distance between your eyes. What you do is measure where stars are in the sky — and then, six months later, when Earth is on the other side of the sun, measure them again.

There's nothing new about this — Baden-Powell wrote about it in Scouting for Boys, the book that launched the Scouting movement, and he had learned it with the British Army in South Africa when soldiers used parallax to measure how far away a (possibly hostile) location was. It worked for them and it works for astronomers — but only for stars up to about 100 light years away. For distances beyond that, the distance between the Earth at one moment and the Earth six months later is too small. And that's a sobering thought, because at 100 light years we haven't even got out of our own galaxy.

The next rung on the cosmic distance ladder is main sequence fitting, which relies on understanding the way stars change as they age (yes, just like humans). But not all stars — only stars called main sequence stars, which means stars of a particular size.

One of the things main sequence stars do as they grow older is to become more red. We also have a rule that says that stars of the same mass and the same age would, if they were at the same distance from us, have the same brightness. (If that rule should ever be shown to be wrong, it would be back to the drawing board for the cosmic distance ladder). Main sequence fitting involves measuring a main sequence star's degree of redness as well as its brightness and comparing these with stars closer to us whose distance we are confident we know accurately thanks to parallax.

Using parallax requires that there be enough light to analyse. It therefore does not work when a star's light, as seen from Earth, is very weak, and that is the situation with stars several million light years from here — which is when the next rung on the cosmic distance ladder comes into play, though it's actually only half a step up because it still uses the principle of main sequence fitting.

This rung uses a class of stars called Cepheid variables. The brightness of Cepheid variables varies in a regular, "pulsating" manner. What makes them valuable for calculating distance is that the time they take to pulsate varies according to their underlying, or intrinsic, brightness: a brighter Cepheid variable will pulsate over a longer period of time than one that is less bright. It's fairly easy to measure the time it takes any particular Cepheid variable to pulsate, which means it's fairly easy to estimate its intrinsic brightness. Comparing that intrinsic brightness with how bright it appears to observers on Earth allows main sequence fitting to be used to estimate its distance from Earth.

And now, perhaps the time is right to admit what would soon become apparent in any case: that some of these measurements we expressed with great confidence are in fact less than 100% certain. One of the twentieth century's great astronomers, Edwin Hubble, used Cepheid variables in the Andromeda galaxy and, on the basis of those variables, decided that Andromeda was slightly less than 1 million light years distant from us. Not very far at all, in cosmological terms. And now, using much the same information as Hubble had, astronomers believe that the Andromeda galaxy is actually about 2½ million light years away. 2.54 million light years, to be exact – but exact is exactly what we can't be, because that figure was arrived at by taking a range of current estimates of Andromeda's distance, all made by skilled and reputed astronomers, and averaging them.

As well as Cepheid variables, Hubble used Type IA supernovas, which are exploding white dwarf stars. They are visible in galaxies at a distance of billions of light years and, because we can calculate how bright the explosions are, we can use the same approach as we use with Cepheid variables to calculate how far away they are. And if it isn't by now obvious that, when we say "calculate" in this context, what we actually mean is "estimate" or, to be entirely frank, "make what we hope will be a pretty accurate educated guess," then perhaps it should be.

When it comes to really extreme distances, we take one more step up on the cosmic distance ladder and reach the rung marked Redshift. Redshift is the Doppler Effect, applied to light from very distant sources instead of the sound of an ambulance passing on the street outside. As an ambulance gets closer, the sound of the siren goes up; as it passes and moves away from you, the sound of the siren begins to fall again. That is sound waves. Light waves act in much the same way. And the way to detect the change, in the case of light waves, is by breaking the light from a distant star or other body into a spectrum (you can do that easily by passing it through a prism). Some parts of the spectrum will contain dark lines where the colours that should show have been absorbed by elements either in the light source or close to it. So far so good, but what really matters is that, the further the light source is from us, the more the dark lines will have moved in the spectrum towards the spectrum's red end. That is redshift. And now we're back with Maurice Chevalier, because what contributes to redshift is not just the fact that the light source is already a very long way away, but that it is moving ever further away at great speed.

We've said several times in this book already that the universe has been expanding ever since the Big Bang. It's still doing so. If we didn't already know that the universe was expanding, the redshift would make it clear. Every galaxy is moving away from every other galaxy. And space-time is being stretched, which means that wavelengths, although in fact they are unchanged, appear to be longer than they are. All of which is helpful in arriving at an estimate of the size of the universe, because the greater the red shift, the further away the source of that light will be.

At this point, we really need drum rolls and crashing cymbals, because we are coming close to the end of this question of the size of the universe. The light with the biggest redshift that we can detect indicates that the galaxies from which that light reaches us left there 13.8 billion years ago. And that might encourage us to believe that the distance from here to the furthest part of the universe is 13.8 billion light years. And since it moves in all directions, that would make the diameter of the universe 27.6 billion light years. But hold the drum rolls and the cymbals, because it isn't quite that simple. In common with all other galaxies in the universe, those galaxies that were 13.8 billion light years away when the light we see now began its journey towards us have gone on moving away from us thanks to the expansion of the universe. And, the further away a galaxy is, the faster it is moving away. The best calculations we can make right now say that, while the observable universe may (as we've just said) be 27.6 billion light years across, and the edge of the observable universe is 13.8 billion light years away in any direction, those galaxies that emitted that light have gone on moving away from us at great speed and have now reached a distance from us of 46.5 billion light years. Double that and we get a diameter of the universe of 93 billion light years. A great deal of it, as we have just shown, is outside the observable universe, and the diameter is increasing daily.

So, the answers to our questions are:

- The diameter of the Milky Way Galaxy (our Galaxy) is about 150 million light years – give or take

- The diameter of the observable universe is about 27.6 billion light years – give or take

- The diameter of the universe as a whole is about 93 billion light years – give or take, and bearing in mind that it's getting bigger all the time

Let's pause for a moment and take in the stunning magnitude of what we've just said. The area of a circle is, as we all learned in school, pi times the square of the radius. But the Universe is not flat, so areas don't enter into it. We don't actually know what shape the Universe is, but if we assume it looks something like a sphere, then the formula to calculate its volume is:

Volume = 4/3 πr^3

Without going through all the troublesome mathematics, that means the observable universe is something like 2.6% of the universe as a whole. Now here's a thought. The SETI Institute calculates that, just in that part of the Universe that we can see, there are 150 billion galaxies other than ours, with an uncountable number of stars with habitable zones and habitable planets in those zones. And we're claiming that the fact that we can't hear or see any sign of an advanced civilisation means we can be certain no advanced civilisation exists? Get outta here!

Of course, all those rungs we've just worked through on the cosmic distance ladder are connected with all the other rungs. So, if one of them is not correct...

IS THE UNIVERSE FINITE OR INFINITE?

Like so many other things to do with cosmology, the answer to this question is: we can make informed guesses, but we don't really know. One theory – and it's a popular theory right now – is that the universe is expanding and will go on expanding for ever. A universe that expands forever is, by definition, an infinite universe. But there is another theory that says that the universe will reach a size beyond which it can no longer sustain expansion and will then start to contract. It may even contract to the point where it forms another enormous black hole, resulting in another Big Bang. According to this theory, there may have been more than one Big Bang in the past, and Big Bangs may be destined to continue into an unknown and unknowable future.

Nor do we know what shape the universe is. In the last section of this chapter we discussed the proportion of the whole universe that we can see – the observable universe – and we discussed that proportion on the basis that the universe is shaped like a sphere. And it may be. It may also be shaped like a doughnut (though scientists prefer the word "torus").

At this point, it's probably as well to mention that astronomers often refer to a "flat" universe and flat in this sense does not mean flat like a football field. "Flat" is used to mean that the universe follows Euclidean geometry: parallel lines never meet, and if you add up all the angles of a triangle you get 180 degrees. This construction enables theories to be expressed in an understandable way.

THE COSMIC MICROWAVE BACKGROUND (CMB)

The European Space Agency launched an observatory into space to study the universe at wavelengths corresponding to infrared, microwave and high-frequency radio. The observatory is called Planck and its purpose is to study the radiation left over from the Big Bang. This radiation has been given the name Cosmic Microwave Background. The universe as it existed immediately after the Big Bang was a hot plasma of particles most of which were electrons, neutrons and protons – but also photons, which carry light. The photons and the free electrons interacted, preventing the photons from travelling far with the result that the universe looked foggy – if there had been anyone to look at it, which there wasn't, it would have seemed opaque.

As the universe expanded, it cooled, because heat comes from energy, the amount of energy was fixed, and the fact that the universe was becoming larger meant that the energy was more widely distributed. When the temperature dropped to about 2700° Celsius, hydrogen atoms were formed from the combination of protons and electrons, which meant that photons were now free – "Have light, will travel." Since then, the universe has expanded greatly, the temperature has dropped hugely, and the photons' wavelengths have become larger to create the background glow seen by radio and far-infrared telescopes. (The glow would be hard to miss – there are some 400 photons in every cubic centimetre of space).

What this cosmic microwave background allows us to do is to see the material from which all of today's galaxies and the stars in them were formed. As has happened so many times, the cosmic microwave background was found by people who were looking for something else; they found it in 1964 and were awarded the Nobel Prize for Physics in 1978.

Planck is the most sensitive of the instruments that have been used to explore the CMB and at some point it should provide enough data either to prove or to disprove the standard model of cosmology.

THE STANDARD MODEL OF COSMOLOGY

What the standard model of cosmology says is, in effect, that the properties of the universe are the same wherever in the universe we happen to be speaking about, though scientists prefer to describe this as a universe that is both homogeneous and isotropic. No direction in space has preference over any other.

At first sight, that does not fit very well with the fact that, wherever we look in the universe, we see structures (stars, galaxies, you name it) that are not only different but possibly unique, but supporters of the standard model say that this is a false analogy. All of those different structures were caused by what in the beginning were tiny fluctuations and differences embedded in the universe as it expanded. The standard model is well supported by observations of many different processes, all of which we can see and verify – but that support only exists if we accept the existence of two other forces that we cannot see and verify: dark matter, and dark energy. Dark matter and dark energy are concepts developed by cosmologists to explain what cannot otherwise be explained.

The history of science contains many examples of theories that were developed to support what could be observed and subsequently were proved by experiment to be correct. The history of science also contains a number of examples of theories that were developed to support what could be observed and subsequently were proved by experiment to be wrong. One of Planck's important tasks is to produce the data that will either confirm or refute the idea that the fluctuations we see in the universe are the result of dark matter and dark energy.

Unfortunately, even if Planck is successful in providing the most detailed understanding yet of the CMB, it still won`t establish for certain whether the universe is finite or infinite. What it may be able to do is to show us more about the shape of the universe and, in particular, whether it is shaped like a doughnut. Light travelling through a torus (doughnut) can either travel around the sides or move in a straight line. In other words, light in a torus-shaped universe would have two ways of reaching the same point. Planck has the potential to tell us whether or not that is happening.

DARK MATTER AND DARK ENERGY

It isn't only in the matter of distance that the bulk of the universe is hidden from us. Even when we take that small portion of the universe that we can see, we can actually only see about 5% of it.

What we can see is known as baryonic matter. The other 95% of the visible universe isn't visible at all – 25% is dark matter, which cannot be seen, and 70% is dark energy, a force that repels gravity but that cannot be felt. Cannot be seen and cannot be felt, that is, by us.

If we can't see it and we can't feel it, how can we possibly know it's there? The answer is that, though we can't see dark matter and can't feel dark energy, we can observe their effects on galaxies and clusters of galaxies. They must exist, because of what they do. If they don't, we have to accept that the physics we use to understand the universe is flawed.

As an example, most of the visible matter in a galaxy is towards the galaxy's centre, And the laws of physics as we understand them mean that, in the case of a spiral galaxy that is spinning, the stars of the galaxy's edge should be moving much more slowly than those near the centre. And they don't. Stars orbit at roughly the same speed, no matter where in the galaxy they may be. The only way at present that we have of understanding this is to assume that something surrounds the galaxy in a sort of halo and that it has a force of gravity to which the stars on the galaxy's edge are subject. That something has been given the name dark matter and its gravitational force has been given the name dark energy.

Other phenomena add force to the argument for the existence of dark matter and dark energy. When galaxies are photographed, the pictures often show strange patterns of light for which no obvious explanation exists unless we accept that it is the light from more distant galaxies and that it is being distorted as it passes through the nearer galaxies by something called "gravitational lensing." The currently accepted explanation for that distortion is that the light is passing through huge clouds of matter that we can't see. Dark matter.

There are other theories about dark matter, and some of them require significant changes to our theory of gravity. (Steven Hawking was very involved in this theorising). According to these theories, gravity has a number of different forms, and the gravity that galaxies are subject to is not the same as the gravity that stops humans from rising from the surface of the earth and spinning away into space. Or, for that matter, that caused an apple to drop on Isaac Newton.

Although it meets the details of a theory expounded by Einstein, dark energy has only been known about since the 1990s. The theory had always been that gravity would slow the universe's expansion, but attempts to measure the slowing down produce the opposite result. Expansion is accelerating, not reducing. It shouldn't happen. But it does – and when you see that what should happen according to an existing theory is the opposite of what does happen, you can't change what is happening so you must change the theory.

And the theory now is that space is not as empty as it seems to be but is filled with a repellent force caused by quantum fluctuations in space and that, as the universe expands, the repellent force grows stronger.

THE FACT THAT WE CAN'T SEE IT DOESN'T MEAN IT ISN'T THERE

And so we have to conclude that it's entirely possible that other advanced civilisations exist, but that they are simply so far away, however we choose to define distance, that we can't see them. And we may never be able to see them. Alien civilisations with very advanced technology may exist but may be so far away that no meaningful two-way communication is possible. We can go further than that and say that, where two such civilisations are separated by thousands of light years, one or both of them may become extinct before such communication can be established. Even if one day we find that they have existed, it may be impossible to contact them either because they are no longer there or because they are simply too far away.

There is a tantalising similarity here between our civilisation and those that have gone before us on Earth. Anyone who has stood by the pyramids and felt that inexplicable sense of mystery can invest a lifetime in decoding and understanding the symbols left by the ancient Egyptians. But talk to them? Ask them questions? It isn't possible, because they're all dead. And analysis of DNA tells us that we can't even communicate with their descendants, because the people who live in Egypt today are not descended from the pharaohs and the people the pharaohs ruled. The genome tells us that the pharaohs had their origin somewhere in Europe. Egypt's population today is largely African.

And if we can't communicate with a civilisation on our own planet because it is distant in time by only a few thousand years, how can we ever expect to make meaningful contact with aliens separated from us by a distance in time millions of times greater than that?

They may well be there. But we can't see them, and we never will. The universe is simply too big.

CHAPTER 12

ISOLATIONISM

A great deal of this book so far has looked at the question, "Where are the advanced civilisations?" and said, "They don't seem to be looking for us, so they probably don't exist." And that is a very self-centred approach. Why should they be looking for us? They may find us very boring. Unless they are bent on universal domination and want to enslave the human race, it's quite possible that they see nothing on Earth that they want to participate in. Nothing fascinates an earthling as much as the earthling him or herself, but it may be unwise to assume that every alien shares that fascination. Or, indeed, that any of them do.

There is a parlour game popular with a certain kind of magazine in which people are asked who they'd like to have dinner with. Honest respondents, depending on both their gender and their orientation, will say things like "Brigitte Bardot" and "George Clooney". For most people, though, this is an opportunity for virtue signalling and putative dinner guests regularly trotted out include people like Shakespeare, the Buddha, Leonardo da Vinci and Dante. Now suppose (and there's going to be more than a little supposing in this chapter) that this game is being played by a Kardashev Scale Level III civilisation's magazine. What a shock it would be for people on earth to discover that the denizens of that civilisation could list a thousand people from right across the universe with whom they would like to break bread – and not a single one of them was from Earth! We are not dinner guest material for advanced civilisation people. They might find us too aggressive. Too sure of ourselves. Or simply too stupid and ignorant. And so they isolate themselves from us. They know we're here, but we don't have anything they want, and they know that if they let us make contact, they're going to be bored. And they hate being bored.

That assumption mentioned in the first paragraph – that if another civilisation is not looking for us, it probably doesn't exist – rests on the idea that what a civilisation will do, once it has reached a certain level, is look for other civilisations in other parts of space. But there we are again, making assumptions about what alien civilisations would do based on what we do. We are looking for other civilisations – therefore other civilisations must be doing the same thing. A lot of very doubtful theorising is based on the idea that there is a universal rule that human instincts will be reproduced throughout the universe. Of course, they may be. But it's just as likely that they are not. To take this particular assumption – that what an advanced civilisation automatically does is to turn outwards and look for others – it may be the exact opposite of the truth. Perhaps when a civilisation reaches a certain level of development, instead of turning outwards it turns inwards.

You want to see that in practice? Without leaving this planet? Visualise two teenagers in a home sufficiently comfortable financially that they have no need to take part-time jobs. See them side-by-side, soft drinks and corn chips to hand, so absorbed in the electronic device each has in his or her hand that they are hardly aware that anyone is beside them. Now see the family that walks into a restaurant for Sunday lunch. Mother, father, grandparents, children. Each child takes out a smart phone and starts pressing keys. Their parents have half an understanding of their children, because they were still at the developmental stage when the earliest electronic devices became available. The grandparents, on the other hand, are bewildered. How is it possible for people to sit at the table with others and not join in general conversation? How can anyone be so absorbed in some make-believe world embodied by and embedded in something not much larger than the box in which Grandma keeps her false teeth that they are unwilling to engage with the people who surround them? Their own family? Their flesh and blood?

Now imagine the same sort of scenes, but this time in a civilisation on a higher level of the Kardashev Scale. No-one now needs to work for a living, because all necessary activities, all production, delivery and maintenance, all agriculture, are roboticised and automated. What does that leave for the aliens to do? Do they just stay home and vegetate? Lie on a chaise longue, sipping wine and eating grapes as Hollywood likes to picture Romans in the time of Nero? Does anyone really imagine that the Romans would have conquered most of the known world and put the fear of God into the rest if they'd really behaved that way? Of course not. And nor is it likely that our imaginary aliens would be so indolent.

Come back to those young people on their iPads and smartphones and laptops and whatever else they play games on. See them in virtual reality headsets, playing games and acting out scenarios and just imagining. Being in the moment and the place, whenever the moment is and wherever the place may be. It's so real – it seems so real – that even today, when there is still a living to be earned, rent to be paid, groceries to be bought, it can be totally absorbing. Simply move someone to another plane, another level, and have them believe that the fantasy they are experiencing is real life.

In civilisations at Kardashev Level II, Kardashev Level III, is it likely that people will still be enjoying that kind of entertainment at that level of quasi-reality? Of course it isn't. The entertainment available to people then will be infinitely richer and more rewarding than anything we have today. (We keep using that word "people," because it's the easiest thing to do. They won't be people in the sense that we are people, you are people – but we don't actually know what they will be (see Chapter 1), so calling them people – while it isn't meant to pre-empt any discoveries – is simply convenient). The people we're talking about will be able to relive history, and not just on their own planet, and everything they see and experience will seem totally real. They'll be able to experience some of the great dramas and operas of their civilisation's cultural history, not by sitting in the stalls but by being on stage and taking part. Imagine being Brutus and stabbing Caesar. Or Juliet, little more than a child at the time, experiencing a level of feeling – of intimacy – of closeness and love – that most people never know however long they live. And not just watching it, hearing it, reading about it – doing it. *Being* it.

People with all of this at their disposal are going to take time out to explain to the people of Earth that it really isn't a good idea to have all these nuclear weapons lying around? No, we don't think so, either.

CYBERMEN

Let's go further. We've just been talking about aliens imagining themselves in other lives. But suppose it isn't imagining? Suppose they've been able to upload themselves into another being? Perhaps a robotic being. Cyberman.

Conversation about the Kardashev Scale sometimes seems very limited. Everything turns on energy – Kardashev defined the very level at which a civilisation stands in his scale in terms of the amount of energy they used and how they got it. (Building Dyson Spheres around stars, for example). And, of course, energy is very important. It's important to all of us. If we couldn't press the switch and turn on the light, keep food fresh and healthy for longer, toast bread without having to sit in front of a fire with a slice on a fork, run a shower or make a cup of coffee without setting a fire first, we'd complain quickly and bitterly. But those things – being able to do those things – that isn't what defines us. Is it? Surely what defines us is what goes on inside our heads.

And so we face the possibility that advanced beings may simply lose interest in their immediate surroundings. In the case of the teenagers here on earth that we've talked about, this lack of interest has to be discarded. They grow up. Fall in love, marry and have children. They need to work for a living in order to pay for those things. For a great many of them, the only time of the year when they feel fully alive will be the two weeks holiday they take in some sunlit, exotic resort. The rest of the time is a day-to-day struggle to keep a roof over their heads, to clothe and feed their children, and make sure those same children get an education so that the cycle can continue into the future. They may not like the baton, but they know they have to hand it to the next generation.

In the case of the Kardashev Level III civilisation, though, that struggle is over. Everything has been taken care of. They will have clothes to wear and food to eat, even if they spend the whole of every day dreaming and never, as the British say, do a hand's turn. Now they can feel fully alive every day, because every day can be a holiday in a sunlit, exotic resort. And they don't need to board an aircraft or a cruise ship to get there.

ACTIVE SETI, OR METI, AND THE SETI PARADOX

A great deal of this book so far has been about SETI, which is the *search* for extraterrestrial intelligence, though we have spoken also about *messaging* extraterrestrial intelligence, or METI (also known as Active SETI). The bulk of Active SETI on earth right now takes place through radio signals, though there are other ways by which people on earth attempt to attract the attention of alien civilisations. Putting a message on the outside of a probe hurled into interstellar space is one. Another which has had occasional popularity resembles the way people bury collections of items under buildings before construction starts. In the case of a building, the collection is usually known as a time capsule and is meant to show people of the future – presumably when the building is knocked down, years or centuries from now – what life was like in this place at this time. They can include pictures of people (to show styles of dress), cars, perhaps a recording of music – the range of possibilities is enormous. With a space capsule, the objects to be blasted into space are chosen less with the aim of showing people descended from us how life was in our time and more to show people from an entirely alien civilisation how life is lived on Earth.

Is this a good idea? There are conflicting views. Aleksandr Leonidovich Zaitsev is a Russian astronomer and radio engineer who takes an optimistic view and describes messaging alien civilisations as a way of overcoming the universe's Great Silence and telling others that they are not alone. (Aleksandr Zaitsev is a common Russian name and it is important to use the Leonidovich to indicate that we are not talking about the ice skating champion, the chess player or a number of others). He describes METI as an unselfish and global action in contrast with the "local and lucrative" nature of SETI. Not everyone agrees.

A possible drawback to enclosing a picture of a human being is that (and this is true whatever the gender of the person in the photograph) it may be received in the same way as pictures reputed at one time to have been shown to foreign soldiers and sailors in Port Said that carried the message (which might also be spoken), "You want to meet my sister?" ("Meet" may be regarded as a euphemism). The question here would be: how to react if an alien lands in a spaceship, produces the photograph and expresses a wish to accept the invitation?

Rather more serious is the possibility that an advanced civilisation that had previously been unaware of Earth's existence – or at least of the fact that people lived here – might, on noting the low Kardashev Scale level to which Earth has so far progressed, decide that here is an opportunity for colonisation. There have even been suggestions that carnivorous aliens might look at the pictures of *Homo Sapiens* in the same way as early hominids looked at the wild, free-ranging animals of their day and think: "Food!" with the result that Earth becomes a large ranch with humans as the livestock.

The SETI Paradox describes the balance between, on the one hand, the very widespread desire to find that advanced extraterrestrial civilisations are trying to contact us and on the other hand, an equally widespread (though by no means shared by all) reluctance to send messages to advanced extraterrestrial civilisations in case doing so ends in tears.

THE ZOO HYPOTHESIS

This brings us to the zoo hypothesis, which takes us back to the beginning of this chapter and the idea that it is their ethical standards that prevent other civilisations from contacting us. There is some support for this idea in the way that research here on Earth has developed. Until very recently, students at universities in the developed West saw no ethical issues in anthropological studies of any sort they chose to conduct in any country, no matter how backward in development. Today, no research student almost anywhere in the developed world will be allowed to carry out a research study of any description without obtaining the approval of the relevant university's Ethics Board and demonstrating that the interests of the subjects of the research are fully protected. That, in essence, is at the heart of the main (though not the only) interpretation of the zoo hypothesis.

The hypothesis says that alien civilisations more developed than ours avoid communicating with Earth in order to allow civilisation here to evolve as it is going to with no contamination from outside. They may have adopted certain standards and set levels of technology and/or political and ethical development that the inhabitants of Earth must attain before they receive contact from another civilisation. In other words, they withhold contact for our sake. There are, though, other possibilities.

One is that the contamination they fear is not from them to us but from us to them. The history of the Earth from its very beginning has been tribal. The men and women of one tribe have seen the men and women of other tribes as alien and, to them, "alien" meant "fit to be killed." As tribes coalesced into nations, the same thing happened on a larger scale. And technological development has not improved things (or so it might be thought); instead, it has meant that the stronger tribes have been able to enslave and slaughter the weaker on a scale not seen before. One can understand an alien civilisation more inclined to peace to decide that warlike and aggressive planets like Earth should be kept in isolation, rather as the United States tried to remain isolated from Europe in the early days of WWI and WWII. (More of this later in the chapter).

Another possibility rests on the assumption that there are many alien cultures and that they respect independent evolution of civilisations either for its own sake or because they see long-running value in a diversity of types of civilisation. In such a case, for an advanced civilisation to share technology and ideas with a backward civilisation (and backward is what, to the kind of alien culture we are talking about, Earth would seem) is to reduce the potential long-run total of diversity in the universe and thereby, over the longer term, reduce the total amount of intelligence the universe is home to.

And a third possibility is the one we began this chapter with: that beings in a Level II or Level III civilisation have simply moved on so far that they are totally absorbed in themselves and more interested in what is happening in the virtual reality they inhabit than in anything earthlings might have to tell them. (If that is the case, we had better assume that they will have made sure before departing for their make-believe world that the AI robots in charge of day-to-day life have also been instructed to deal with invasions from other advanced civilisations that have not decided that virtual reality is better than real life and are bent on colonising them).

THE ZOO HYPOTHESIS REQUIRES A FEDERATION

It is easiest to believe in the possibilities we have just outlined if we also believe that advanced extraterrestrial civilisations are joined together in some sort of federation and accept direction from a single council or even a single ruler. The reason that is important is that, if there are indeed a number of advanced civilisations out there, and if there is no unifying force to hold each one of them to the same policy, then it would only take one of those civilisations with different ideas to end the possibility of isolating Earth.

And there is also the question: what happens if we force our attentions on them? It's one thing for more advanced civilisations to ignore Earth while Earth knows nothing about them and is unable to make contact. It's a different matter entirely if we land a spacecraft there – or, even more so, send digging equipment and start mining their minerals. Would they simply ignore that? Would we?

TESTING THE ZOO HYPOTHESIS

Researchers construct hypotheses in order to test them. Testing the zoo hypothesis presents challenges, not the least of which is: if we are not being isolated by other civilisations because no other civilisations exist, no way of testing that hypothesis (the non-existence of other civilisations) with certainty has yet been devised. Nevertheless, success is more likely to come as a result of organised, structured testing than in response to random transmissions.

Let's start with radio transmissions, because those are the most frequently suggested ideas. The most immediate problem is easy to understand; just imagine that you are a monoglot English speaker and a man moves in next door who speaks nothing but Arabic. You're out in the garden raking up leaves and the neighbour stops by and speaks to you. You know he's trying to give you some sort of message (even if it's just "Hi"), because you can hear his voice and see his lips move. But what *is* the message? What is he saying? You don't know, because you don't understand the protocols by which the message is constructed. (In this case, the protocols are the language being used). It would be much the same if we transmitted into space a radio message saying – in English – "Hello. Earth calling." If this message happened to be picked up by radio receivers in some distant galaxy, whoever the receiver belonged to would know that they had received a signal of some sort. It's possible that the way the signal was structured would tell them that it was meant to be a message. But, unless they spoke English (a somewhat unlikely proposition), how could they possibly know what was being said?

The expression used for this sort of radio message is IRM (Interstellar Radio Message). If we received one, then detecting it would be one thing and extracting its message quite another. The same would be true for any distant civilisation receiving an IRM from us. Extracting the message is difficult because we know nothing about the language (the protocol) in which it was transmitted.

The conclusion must be that, if we want to send a message into space and have any hope that a civilisation in another galaxy could not only receive it (which is itself the longest of long shots) but also understand what it said then we have to give some thought to structuring the message in such a way that whoever received it would have some chance of decoding it. The message, then, would need to have two components: first, a "hailing component" that clearly says, "This is a message from a sentient being" and second, the content of the message must be encoded in a way that will make understanding at least possible. It should be the least secure type of message that it is capable of imagining.

Michael W Busch of UCLA and the SETI Institute created a binary language for use as a hailing component. It was used in the Lone Signal project set up to try to communicate by radio message which (and this was through no fault of Michael W Busch) ended soon after it had begun because it failed to raise enough money from its crowdfunding appeal. But at the distance at which a message will be received, recognising it for what it is might prove more difficult than we imagine. All radio signals have a degree of modulation – that's how they work – but the modulation inflicted on a radio signal sent over intergalactic distances would change it to the extent that it could be entirely unrecognisable. There is a degree of degradation, however slight, in signals received here from spacecraft approaching the edge of our own solar system. A signal sent over a distance of several galaxies would be far more degraded; one to or from the far reaches of the Universe might well be indistinguishable from cosmic noise.

Theorists suggest that messages containing ideograms might hold their structure better and over greater distances than those containing text and this would solve two problems: the degradation question we've just been discussing, and the matter of language. It would still be necessary, though, to find an ideogram that represented something in a way likely to be capable of being interpreted and understood by an alien civilisation. Look how long it took to interpret ideograms from Ancient Egypt. And they were put there by members of the same species as us!

Just what are the risks?

METI is the transmission of messages into outer space in the hope that they will be received and decoded by advanced civilisations. We've asked this question before, but it is sufficiently important to ask it again: is this wise? There are those who say it is and those who say it isn't. The same Aleksandr Leonidovich Zaitsev we have already mentioned had no doubts. According to him, the risks of not reaching out to other civilisations are greater than the risks of doing so. Not everyone agrees, and Steven Hawking was one of those who did not. We could be exposing ourselves to the risk of colonisation. Worse: we could be exposing ourselves to the likelihood of being wiped out.

Suppose the inhabitants of another planet where the civilisation is much more advanced than ours look at Earth and think, "That looks like a nice place to live. Let's discover it." If they decided to do so, would they allow us – the people already here – to go on living in the way that we have without being disturbed, colonised or liquidated? Possibly. It might be interesting, though, to collect the views on that question of the Caribs in the Caribbean. Actually, however interesting it might be, it wouldn't be possible because the Caribs were indeed liquidated. Okay, then: ask the native peoples of the USA and Canada. Did they survive colonisation without harm befalling any of them? Survive without their native culture being sidelined? Survive without being forced to adapt to the ways of the colonists who, after all, were stronger than them, had better weapons, and were fortified by the certainty that their God wanted them to bring the colonised people into order? Survive without being corralled in reservations, having their hunting grounds expropriated, and seeing the death of a large number of their fellows? You know the answer to those questions.

All of that assumes that aliens would be ill disposed towards us, and warlike in their own nature. They might be They might be exactly the opposite. We don't know.

A great many astronomers, philosophers, politicians and ordinary people have come together on a number of occasions to work out what steps people on Earth should take to agree between themselves whether to transmit messages to advanced civilisations that might turn out to be hostile. It's nice to know that those conversations are being had, but none of them has prevented messages from being sent. They are a bit like the Geneva Convention and treaties on world trade: observed when observing them is in the interests of the observer and ignored otherwise.

Others, like Liverpool academic João Pedro de Magalhães, take no position on the question of whether messaging advanced civilisations in other galaxies is dangerous or not; in their view, the danger will be neither greater nor less than it already is if any advanced extraterrestrial civilisation is both aware of our existence and capable of reaching the solar system. They will want to do us harm or they won't; either way, there won't be much we can do about it.

IS IT EVEN POSSIBLE?

There has been talk of immensely powerful beacons for interstellar transmissions, and some astronomers have dampened down discussion along these lines on the basis that the energy required would be immense, that it is energy access and use that determines where a civilisation sits on the Kardashev Scale, and that we should not forget that we haven't yet even reached Kardashev Level I. Put simply, we couldn't do it.

But that assumes a particular kind of beacon. The ordinary consumer electronics we have now are capable of producing transmitters that can beam a signal to stars that are not desperately far away, and progress in consumer electronics is extraordinarily rapid. It's also true that, once we know where the civilisation we want to talk to is located, the signal can be highly directional and that would require a great deal less energy.

IN ANY CASE, IS ISOLATION POSSIBLE?

It may be that the whole conversation is pointless. The isolation we discuss is two-way: we've raised the possibility that advanced alien civilisations may have placed Earth in quarantine and be adopting an isolationist policy towards us and we've also raised the possibility that it might be wise for us to adopt an isolationist policy of our own in order not to attract visitors who might not have our best interests at heart and of whom it may prove impossible to rid ourselves.

But is there any choice? Is isolation possible? Can it work?

Indications from our experience here on Earth are not hopeful. America pursued an isolationist policy on the outbreak of World War I and World War II. America ended up fighting in both of those wars and losing a great many of its citizens, to say nothing of transforming its political make-up out of all recognition. Ireland adopted a policy of neutrality during World War II, but there was no shortage of Germans in Ireland looking for ways to use the country as a platform for invading Britain, and no shortage of Brits hunting them. Switzerland, Spain and Portugal all had isolationist policies during that war, and all of them were able to avoid being embroiled in conflict, but the price was high and staying out of the war did not mean not having enemy combatants on their soil.

And you can only remain in isolation if others allow it. When America was attacked by Japan, its isolation was forcibly ended.

So what conclusion should we draw? Probably this: that there is very little point in talking about isolation at present but that, if First Contact ever becomes a reality – that is, if we on Earth find ourselves in any kind of communication with an advanced extraterrestrial civilisation – the debate will be renewed with vigour. There will be an isolationist camp which will insist that it alone is right and that anyone who disagrees is an idiot. There will be a reaching-out camp which will insist that its way is the only way, and that anyone who disagrees is a moron.

Political business on earth as usual, in fact.

Chapter 13

Break on Through – to the Other Side

Some of the most amazing engineering work in the world today is not exemplified by the buildings going up in the Arabian Gulf, in London and in New York – stunning though some of the buildings in Dubai and Qatar are. The most astonishing developments in engineering, in technology and in science are also those with the greatest promise for the future and they are happening at the atomic and molecular level and not at the size of tall buildings. We are talking about nanotechnology and nanoengineering. Changing materials at the level of the atom and the molecule also changes the material's properties. Nano-engineered concrete is able to make structures that are far more durable, less liable to crack, easier to work and use less water. And concrete is among the materials experiencing less stunning advances than some others.

When we begin to think about how small things can get, and the impact working on very small scales can have, it isn't long before we ask the question: what's the smallest thing there is? And, so far as we can tell, the answer to that is: the event horizon at the centre of a black hole.

We've talked about black holes before in this book, and now we're talking about them again, and the reason is that black holes may well be where our future lies.

It's easy to read that statement and shiver. We've read sci-fi books and we've seen sci-fi movies in which the black hole is the most dread feature the universe has to offer. In the 1930s and 1940s, boys' comics had lots of stories about divers who inadvertently put a foot into an immense shell-like creature that promptly closed its jaws and held the diver fast. Black holes have held the imagination of later generations with the same terror – fly your spacecraft too close to a black hole, you'll be sucked into it and no one will ever hear from you again because what goes into a black hole never emerges.

One reason why black holes can generate this terror is because we don't know very much about them. We think we do, but the fact is that we think we know more than we do. And it's always easier to be frightened of what we don't know than of what we do. One reason we know less than we think we do is that two of the theories dearest to physicists – the general theory of relativity and the theory of quantum mechanics – can't (quite) be reconciled with each other. Some aspect of one or other of the theories – or both of them – needs to be amended. When the late Steven Hawking died, he was working towards a General Theory of Gravity. Other physicists are still doing so, and it's possible that when that theory emerges it will not only tell us where the discrepancies between general relativity and quantum mechanics are and how to fix them but will also allow us to understand what a black hole really is.

But one thing we can be quite certain of is that black holes contain incredible (to us) amounts of energy. Enough energy to supply the total energy needs of every human being on Earth for trillions upon trillions of years. Long after the sun has died and its light has gone out, Earth could still be powered from the energy in a spinning black hole. And that's the point. Stars die. The Dyson spheres we've talked about up to now have been built around stars. But wouldn't it make far more sense to build them around black holes? Black holes have far more energy than any star and, if their life is finite, then we don't know about it beyond saying that it's entirely possible that this was behind the Big Bang: that a previous universe collapsed into an almost infinitely dense black hole which then exploded to create the replacement universe in which we now live. And if our descendants are drawing all their energy from a black hole, and if the whole universe then collapses into a single black hole, our descendants won't need energy any longer – because they will be down there in the black hole with everything else. And they'll be dead. Waiting for the moment when everything erupts again in indescribable heat and at unimaginable speed. Until then, all energy requirements will be met from the black hole.

Which is a lovely idea – but how on earth are we to get our hands on all that energy?

And the answer to that question is almost certainly that we – in our present form – won't be able to do so. We couldn't survive long enough to build a Dyson sphere around a black hole. But the robots into which (see Chapter 9) our descendants will have morphed, transformed themselves, or simply clambered will be able to do so.

There's more!

Even though we don't know what a black hole is – that is, we don't really know – we know there are some exciting possibilities and questions. A black hole could be the doorway into another universe. Or into a dimension that most humans can't even imagine, let alone visualise. And a reason why we can't see advanced extraterrestrial civilisations may be that they have gone through that door into a different dimension. One through which our descendants may, at some time far, far into the future, follow them. Unless, that is (see Chapter 12) they know all about us and have carefully locked the door to make sure that such savages can't get anywhere near them.

THE TRANSCENSION HYPOTHESIS

We spend a great deal of time talking about outer space when perhaps we should be more concerned with inner space. The transcension hypothesis suggests that the evolution of civilisations as they progress up the levels of the Kardashev Scale leads them into a domain in which everything becomes progressively smaller and more efficient. Everything, in this context, includes not just matter but also energy, time and space. And the end of the development route is what amounts to a black hole.

One of the selling points for the hypothesis is that it explains why we haven't seen any other intelligent civilisations. Or heard from them, for that matter.

Why would something like a black hole attract advanced intelligences? It's not as daft an idea as it may sound. A black hole can be regarded as the most perfect computing and learning environment. It is, as we've already seen, a source of infinite energy. Because of the strange things that happen to all dimensions including time inside a black hole, time travel may well be possible – though there are those who say it would only be possible in a forward direction. The most exciting prospect of all could be the potential for replicating universes (see Big Bang).

And now we can propose a reason for the ethical injunctions on communicating with Earth that Chapter 12 suggested may be in force for higher, more advanced civilisations. Because what do we imagine is going to happen in that black hole into which, should we ever reach the right level of development, we may hope to find ourselves? One likelihood is that civilisations will merge. Another is that there will be natural selection (and we already know what a powerful impetus that has in evolution). We suggested in Chapter 12 that the prohibition on contact with Earth – the isolation into which we said Earth might be placed – could possibly be because of a desire for the greatest possible diversity of civilisation. And here is the justification for that suggestion. If the end object is merger and natural selection, then it must be clear that there is an inbuilt need not to allow any civilisation to be diluted from the outside.

CAN WE EVER KNOW?

Probably not, at least until we develop enough to take our place with the other advanced civilisations in the black hole. But... Perhaps.

Supermassive black holes are surrounded by things called accretion discs. You can't see a black hole; you *can* see the accretion discs. The black hole generates so much gravity that it accelerates the spinning of the particles making up the disc, and therefore of the disc itself, to quite amazing speeds and in doing so releases heat as well as x-rays and gamma rays.

What, then, constitutes an accretion disc? How does it come to be formed? Accretion discs are made up of gas, dust and other material that has got quite close to the black hole but not close enough to fall into it. It is formed into a flattened band of material spinning around the event horizon. We know that. But do we? In fact, that is only a theory that we happen to have right now, and the history of study of the universe going right back to the earliest times is one of continuous abandoning of theories that had seemed to work but no longer do. Perhaps that isn't what an accretion disc is at all. Because one thing we do know about accretion discs is that they are amazingly effective at harvesting energy.

And isn't that what we've been saying that those advanced civilisations are going to need as they head towards transcension?

We don't know for sure that accretion discs were not built by civilisations much more advanced than ours in order to harvest the energy that propelled them up the Kardashev Scale and towards their ultimate destination inside a black hole. They might not have been. There is every possibility that they weren't. But they might have been. Those accretion discs might be the practical implementation of a Dyson Sphere and the only evidence of an advanced extraterrestrial civilisation that we are ever going to see until we reach that level of development ourselves.

We – that is, you and I – will never know, because it will be millions of years before the truth becomes clear to the inhabitants of Earth and you and I will have turned to dust long before that.

But someone from this planet will know. And, if you've been with us all the way since Chapter 1, then you know that that someone won't look the tiniest bit like us. But he or she will still be us. In whatever form we have become.

The next clear night we have, find the least light-polluted place you can and stare out into space at the Milky Way. Isn't it amazing? Doesn't it make us all feel small? And yet, at the same time, immensely potent? It's taken humankind a long time to reach the rather primitive stage we are at now. It will take even longer to get where we're going. But have faith. Because we are going there.

END

EPILOGUE

Writing a book is more than just the creative process. One has to research the competition on the subject and make sure that there are not already a million titles all competing for the same set of eyeballs and index fingers on the <Buy> button. Once the research done, some reflection is in order. What is the target audience? How can you market to them?

My first little victory was getting the support of Nick Pope. Nick's work was always something I found very interesting. His life is fascinating when you peel back all the layers. Nick used to run the British Government's UFO Project and is the world's leading expert on UFOs, the unexplained and conspiracy theories. His works of literature inspired me to begin down the path on the subject of extra-terrestrial life. I wanted to know all the possible reasons why we might not have found alien life yet, and I wanted to flesh them out for you, the reader. Nick has a website where you can read more, and take a look at his books. A simple google search will turn him up.

While researching the subject matter, we reached out to many people that make up the community of believers, non-believers, ancient alien theorists, and scientists that all have the same sense of wonder about the existence of alien life. Everyone's opinion varied on the subject, from people that outright believed in the existence of alien life in our midst, to others that simply think we are the only intelligent life form in the entire universe. But they all wanted to know. They want to believe.

At some point life got in the way and this project was put on the back burner. Our writing projects (and there are quite a few of them) run simultaneously with a real life careers for both of us that drain lots of energy and creativity. Writing doesn't pay the bills yet, so to keep our significant others in the lifestyles that they have become accustomed to, Mark and I have to also work full time jobs.

Recently, Mark broke his right leg in three places. I took my wife to Winnipeg for a weekend to help him and his girlfriend with some cooking and errands to get him settled in for his recovery. I hadn't been back to Winnipeg for about ten years, and was impressed at how the city had changed. Sitting in Mark's front yard, with his leg in a cast, we decided on Sept 18th, 2018 as our release date. It was an ambitious deadline, but as they say, nothing focuses the mind more than a deadline and a self-imposed one even more. Besides, Mark broke his leg, not his hand.

Why Sept 18th? There is really no reason. It's a Tuesday, and it's squarely after everyone is back to school. September is a good month for book sales and we felt that the date was right. We also needed time to get the French version right.

About the French version. Mark and I live in Canada and I live in Montreal, Quebec. My wife and I have a sizable circle of friends and family whose first language is French. After publishing our first book *"History's Greatest Deceptions and Confidence Scams"* I faced the frequent question. "When are you going to put something out in French?"

Our country was founded on the equal partnership of the two solitudes of French and English people getting together, breaking bread, learning about one another and developing tolerances over each other's cultures. Mark and I wanted to simultaneously produce this work in both of Canada's Official languages.

Translating a book and doing it on a shoestring budget is not easy. It helps that I can read French adequately but my writing skills are sorely lacking. Enter Marco Lambert and Daniel Lefaivre. They each agreed to take a part of the book and fix the first two translations we had produced. If it wasn't for these two fine gentlemen, we would have been in deep trouble.

Our energy and effort of over a year and a half has gone into this work. We interviewed many interesting characters and made wonderful friends and connections with collaborators. Each connection is treasured and if some of you are reading this, we thank you for your support and help.

Finally, if you enjoyed this work, let us know. We are on Facebook, and have a website. Mark and I are always looking for Beta readers and Advance reviewers. Sign up on our site and we'll be happy to include you in our ARC club. Oh and go and review us on Amazon and Goodreads. The lifeblood of writers is the review process with Amazon and Goodreads. If you didn't like something in this book, reach out. We can talk about it and make it right.

Thank you all for encouraging us. We've got a few more titles coming out soon. Watch our site and Facebook page, and if you have an interest in history, please see a complimentary section on *"History's Greatest Deceptions and Confidence Scams"* in the bonus material section at the end of this book.

ABOUT THE AUTHORS

About the Authors

Lifelong friends, Mark Rodger and Steven Lazaroff have studied, researched and debated the Fermi Paradox, the possible existence of extraterrestrial life, and the SETI program that is searching it out. Amateur UFOlogists, historians and explorers, they have applied their research skills and passion to attempt to explain, in layman's terms, the hunt for life amongst the stars, and to attempt to explain why Humanity may not yet have had contact with alien life.

Steven Lazaroff lives in Montreal, Quebec.

Mark Rodger is from Winnipeg, Manitoba.

Both are unapologetic Canadians.

INTRODUCTION FOR 'HISTORY'S GREATEST DECEPTIONS AND CONFIDENCE SCAMS'

The term "scam" is really a new word in the English lexicon and

has come to supersede its older and more distinguished original cousin, "the confidence game" or "con game", as it became popularly known at the time. One constant in human history is the tendency to want to shorten and simply some of the most descriptive concepts we have, to anything that can be mumbled as a single syllable mouthful.

Throughout history, there have always been fraudsters and tricksters ready and willing to part people and their money with smooth talking and tall tales, but the first formally recorded "confidence trick" was uniquely American in its origins and set the bar for both simplicity and sheer guts, both hallmarks of the most successful frauds ever perpetrated.

In the late 1840's the east coast of the United States was awash with the nouveau riche, and men wearing top hats to try and look important. Good manners and polite society were everything unless you were a slave in which case the top hat was entirely optional. It was the age of Jane Austen, white gloves, carriages and over-the-top manners. It was also the time of pocket watches, dangling from gold chains. Victorian sensibilities dictated that the bigger and shiner the watch, the bigger and shinier the man.

Enter one William Thompson, arguably the originator of the term "confidence man", a genius operator and a personal hero to the career grifter. Little is known about where he came from, but what is certain is that he had his thumb on the pulse of well-heeled suckers strolling the walkways and avenues of Manhattan in the mid-nineteenth century.

Meeting someone was a rigid, formal affair with protocol and procedures; the handshake, tip of the hat and bow were rigidly choreographed. Failure to introduce oneself properly or be introduced according to accepted custom was seen as an embarrassment to both parties – and embarrassment was worse than a bleeding head wound, to be avoided at all costs. Operating in New York in the 1840s, William was a keen observer of human behaviour.

He realized that with such pomp and ceremony surrounding every introduction, it was considered the ultimate in bad manners not to remember people that one might have been acquainted with – he calculated that when confronted with a stranger that said he was a friend, most men would likely act like they remembered a meeting that had never happened.

William thought he might be able to leverage this, and so would often stroll along the city streets, until he spotted an upper-class sucker, at which time he would approach and pretend to know them and be a past acquaintance, someone that they had met before. Rather than be embarrassed, the mark would usually smile, nod and pretend that he knew who William was – better that than risk dishonour, or a pistol duel – which was how some matters of honour were settled at the time.

After some friendly chatting, and a little trust-gaining, Thompson would throw out his hook, asking "Have you confidence in me to trust me with your watch until tomorrow?" He wasn't all about watches – sometimes he would ask for money. It's good to diversify. More often than not, the mark would part with the watch or the money (or sometimes both) and William would depart, promising to meet them the next day to return the property.

Naturally, he didn't keep the next day's appointment and would often stroll away, laughing to himself.

He repeated this game dozens of times until he had the bad luck to happen across a former victim, who promptly summoned a roving policeman who gave chase. After a frantic foot pursuit through Manhattan and a dramatic struggle, William was bodily subdued and arrested. Perhaps he was slowed down by the weight of all those pocket watches; it was reported that he had several on him at the time he was caught.

His arrest and the subsequent article in the New York Herald called "Arrest of the Confidence Man" made headlines across the country; he was headed to trial in 1849. The press noted his specific appeals to victims' "confidence" and thereafter he was known in the press as "The Confidence Man". And so the term was born, and "confidence game" or "con" became part of our vocabulary, and spawned an endless series of quick-buck fraudster copycats that said, "me too"!

This is the story of some of the greatest.

Printed in Great Britain
by Amazon